ALSO BY ALAN LIGHTMAN

The Accidental Universe

THE
Accidental Universe

THE WORLD YOU THOUGHT YOU KNEW

ALAN LIGHTMAN

PANTHEON BOOKS, NEW YORK

Copyright © 2013 by Alan Lightman

All rights reserved. Published in the United States
by Pantheon Books, a division of Random House LLC, New York,
and in Canada by Random House of Canada Limited, Toronto, Penguin
Random House Companies. Originally published in the United Kingdom by
Corsair, an imprint of Constable & Robinson, London, in 2013.

Pantheon Books and colophon are
registered trademarks of Random House LLC.

Selected chapters in this work were previously published in the following:
"The Spiritual Universe," Part I, in *Salon* (October 2011); "The Temporary
Universe" in *Tin House* (Spring 2012); "The Spiritual Universe," Part II, in *The
Commercial Appeal* (June 2012); "The Accidental Universe" and "The Gargan-
tuan Universe" in *Harper's* (December 2012); and "The Symmetrical Universe"
in *Orion* (March/April 2013).

Library of Congress Cataloging-in-Publication Data
Lightman, Alan P., [date]
The accidental universe : the world you thought you knew / Alan Lightman.
pages   cm
Includes bibliographical references.
ISBN 978-0-307-90858-2
1. Cosmology. 2. Universe. 3. Astronomy—Philosophy.
4. Physics—Philosophy. 5. Intuition. 6. Instinct. 7. Thought and thinking.
8. Knowledge, Theory of. 9. Lightman, Alan P., [date] I. Title.
QB981.L55 2013   523.1—dc23   2012047550

www.pantheonbooks.com

Jacket design by Pablo Delcán
Book design by Robert C. Olsson

Printed in the United States of America
First American Edition
2 4 6 8 9 7 5 3 1

To my dear friends Sam Baker, Alan Brody, John Dermon, Hok Dy, Owen Gingerich, Micah Greenstein, Bob Jaffe, Peter Meszaros, Russ Robb, David Roe, Peter Stoicheff, and Jeff Wieand

CONTENTS

PREFACE

In October 2012, I attended a lecture given by the Dalai Lama in a cavernous auditorium at the Massachusetts Institute of Technology. Even without words, the moment would have been profound: one of the world's spiritual leaders sitting cross-legged in a modern temple of science. Among other things, the Dalai Lama spoke about *śūnyatā*, translated as "emptiness," a central concept in Tibetan Buddhism. According to this doctrine, objects in the physical universe are empty of inherent and independent existence—all meaning attached to them originates in constructions and thoughts in our minds. As a scientist, I firmly believe that atoms and molecules are real (even if mostly empty space) and exist independently of our minds. On the other hand, I have witnessed firsthand how distressed I become when I experience anger or jealousy or insult, all emotional states manufactured by my own mind. The mind is certainly its own cosmos. As Milton wrote in *Paradise Lost,* "It [the mind] can make a heaven of hell or a hell of heaven." In our constant search for meaning in this baffling and temporary existence, trapped as we are within our three pounds of neurons, it is sometimes hard to tell what is real. We often invent what isn't there. Or ignore what is. We try to impose order, both in our minds and in our conceptions of external reality. We try to connect. We try to find truth. We dream and we hope. And underneath all of these strivings, we are haunted by the suspicion that what we see and understand of the world is only a tiny piece of the whole.

Modern science has certainly revealed a hidden cosmos not visible to our senses. For example, we now know that the universe is awash in "colors" of light that cannot be seen with the eye: radio waves and X-rays and more. When the first X-ray telescopes pointed skyward in the early 1970s, we were astonished to discover a whole zoo of astronomical objects previously invisible and unknown. We now know that time is not absolute, that the ticking rate of clocks varies with their relative speed. Such incongruities in the passage of time are unnoticeable to us at the ordinary speeds of our lives but have been confirmed by sensitive instruments. We now know that the instructions for making a human being, or any form of life, are encoded in a helix-shaped molecule found in each microscopic cell of our bodies. Science does not reveal the meaning of our existence, but it does draw back some of the veils.

The word "universe" comes from the Latin *unus,* meaning "one," combined with *versus,* which is the past participle of *vertere,* meaning "to turn." Thus the original and literal meaning of "universe" was "everything turned into one." In the last couple of centuries, the word has been taken to mean the totality of physical reality. In my first essay, "The Accidental Universe," I discuss the possibility that there may exist multiple universes, multiple space-time continuums, some with more than three dimensions. But even if there is only a single space-time continuum, a single "universe," I would argue that there are many universes within our one universe, some visible and some not. Certainly there are many different vantage points. These essays explore some of the views, both the known and the unknown.

The Accidental Universe

The Accidental Universe

In the fifth century BC, the philosopher Democritus proposed that all matter was made of tiny and indivisible atoms, which came in various sizes and textures—some hard and some soft, some smooth and some thorny. But the atoms themselves were accepted as givens, or "first beginnings." In the nineteenth century, scientists discovered that the chemical properties of atoms repeat periodically, as in the so-called Periodic Table, but the origins of such patterns remained mysterious. It wasn't until the twentieth century that scientists learned that the properties of an atom are completely determined by the number and placement of its electrons, the subatomic particles that orbit the nucleus of the atom. These details, in turn, have been explained to high accuracy by modern physics. Finally, we now know that all atoms heavier than helium were created in the nuclear furnaces of stars.

The history of science can, in fact, be viewed as the recasting of phenomena that were once accepted as "givens" as phenomena that can now be understood in terms of fundamental causes and principles. One can add to the list of the fully explained: the hue of the sky, the orbits of planets, the angle of the wake of a boat moving through a lake, the six-sided patterns of snowflakes, the weight of a flying bustard, the temperature of boiling water, the size of raindrops, the circular shape of the sun. All of these phenomena and many more, once thought to have been fixed at the beginning of time or the result of random events thereafter, have ultimately been explained as *necessary* consequences of the fundamental laws of nature—laws found by us human beings.

This appealing and long trend in the history of science may be coming to an end. Dramatic developments in cosmological findings and thought have led some of the world's premier physicists to propose that our universe is only one of an enormous number of universes, with wildly varying properties, and that some of the most basic features of our particular universe are mere *accidents*—random throws of the cosmic dice. In which case, there is no hope of ever explaining these features in terms of fundamental causes and principles.

It is perhaps impossible to say how far apart different universes may be, or whether they exist simultaneously in time. But, as predicted by new theories in physics, the many different universes almost certainly have very different properties. Some may have stars and galaxies like ours. Some may not. Some may be finite in size. Some may be infinite. Some may have five dimensions, or seventeen. Physicists call the totality of universes the "multiverse," a word that sounds as if it came from a Robert Heinlein novel. Physicist Alan Guth, a pioneer in cosmological thought, says: "The multiple universe idea severely limits our hopes to understand the world from fundamental principles." And the philosophical ethos of science is torn from its roots. As put to me recently by the Nobel Prize–winning physicist Steven Weinberg, a man as careful in his words as in his mathematical calculations: "We now find ourselves at a historic fork in the road we travel to understand the laws of nature. If the multiverse idea is correct, the style of fundamental physics will be radically changed."

The scientists most distressed by Weinberg's "fork in the road" are theoretical physicists. Theoretical physics is the deepest and purest branch of science. It is the outpost of science closest to philosophy, and religion. Experimental scientists occupy themselves with observing and measuring the cosmos, finding out what stuff exists, no matter how strange that stuff may be. Theoretical physicists, on the other hand, are not satisfied with observing the universe. They want to know *why*. They want to explain all the properties of the universe in terms of a few fundamental principles and parameters. These fundamental principles, in turn, lead to the "laws of nature," which govern the behavior of all matter and energy. An example of a fundamental principle in physics, first proposed by Galileo in 1632 and extended by Einstein in 1905, is the following: All observers traveling at constant velocity relative to one another should witness identical laws of nature. From this principle, Einstein derived his entire theory of special relativity. An example of a fundamental parameter is the mass of an electron, considered one of the two dozen or so "elementary" particles of nature. As far as physicists are concerned, the fewer the fundamental principles and parameters, the better. The underlying hope and belief of this enterprise has always been that these basic principles are so restrictive that only one self-consistent universe is possible, like a crossword puzzle with only one solution. That one universe would be, of course, the universe we live in. Theoretical physicists are Platonists. Until the last few years, they believed that the entire universe, the one universe, was generated from a few principles of symmetry and mathematical truths, perhaps throwing in a handful of parameters like the mass of the electron. It seemed that we were closing in on a vision of our universe in which everything could be calculated, predicted, and understood.

However, two theories in physics, called "eternal inflation" and "string theory," now indicate that the *same* fundamental principles, from which the laws of nature derive, lead to many *different* self-consistent universes, with many different properties. It is as if you walked into a shoe store, had your feet measured, and found that a size 5 would fit you, a size 8 would also fit, and a size 12 would fit equally well. Such wishy-washy results make theoretical physicists extremely unhappy. Evidently, the fundamental laws of nature do not pin down a single and unique universe. According to the current thinking of many physicists, we are living in one of a vast number of universes. We are living in an accidental universe. We are living in a universe uncalculable by science.

"Back in the 1970s and 1980s," says Alan Guth, "the feeling was that we were so smart, we almost had everything figured out." What physicists had figured out were very accurate theories of three of the four fundamental forces of nature: the strong nuclear force that binds the particles in atomic nuclei together, the weak force that is responsible for certain kinds of radioactive decay, and the electromagnetic force between electrically charged particles. And there were prospects for merging quantum physics with the fourth force, gravity, and thus pulling it into the fold of what physicists called the Theory of Everything. Some called it the Final Theory. These theories of the 1970s and 1980s required the specification of a couple dozen parameters corresponding to the masses of the elementary particles, and another half dozen or so parameters corresponding to the strengths of the fundamental forces. The next logical step would have been to derive (if possible) most of the elementary particle masses in terms of one or two masses, and the strengths of all the fundamental forces in terms of a single fundamental force.

There were good reasons to think that physicists were poised to take this next step. Indeed, since the time of Galileo, physics has been extremely successful in discovering principles and laws that have fewer and fewer free parameters and that are also in close agreement with the observed facts of the world. For example, the observed rotation of the ellipse of the orbit of Mercury, a tiny 0.012 degrees per century, was successfully calculated using the theory of general relativity. And the observed magnetic strength of an electron, 2.002319 magnetons, was accurately derived with the theory of quantum electrodynamics. More than any other science, physics brims with such highly accurate agreements between theory and experiment.

Guth started his physics career in this sunny scientific world. Now sixty-four years old and a professor at MIT, he was in his early thirties when he proposed a major revision to the Big Bang theory, called inflation. We now have a great deal of evidence suggesting that our universe began as a nugget of extremely high density and temperature about fourteen billion years ago and has been expanding, thinning out, and cooling ever since. The theory of inflation proposes that when our universe was only about a trillionth of a trillionth of a trillionth of a second old, a peculiar type of energy caused the cosmos to expand very rapidly. A tiny fraction of a second later, the universe returned to the more leisurely rate of expansion of the standard Big Bang model. Inflation solved a number of outstanding problems in cosmology, such as why the universe appears so homogeneous on large scales.

When I visited Guth in his third-floor office at MIT one cool day in May, I could barely see him above stacks of papers and empty Diet Coke bottles on his desk. More piles of papers and dozens of magazines littered the floor. In fact, a few years ago Guth won a contest sponsored by *The Boston Globe* for the messiest office in the city. The prize, he says, was the service of a professional organizer for one day. "She was actually more a nuisance than a help. She took piles of envelopes from the floor and began sorting them according to size."

Guth is still boyish looking. He wears aviator-style eyeglasses, has kept his hair long since the 1960s, and chain-drinks Diet Cokes. "The reason I went into theoretical physics," Guth tells me, "is that I liked the idea that we could understand everything (i.e., the universe) in terms of mathematics and logic." He gives a bitter laugh. We have been talking about the multiverse.

While challenging the Platonic dream of theoretical physicists, the multiverse idea does explain one aspect of our universe that has unsettled some scientists for years: according to various calculations, if the values of some of the fundamental parameters of our universe were a little larger or a little smaller, life could not have arisen. For example, if the nuclear force were a few percent stronger than it actually is, then all of the hydrogen atoms in the infant universe would have fused with other hydrogen atoms to make helium, and there would have been no hydrogen left. No hydrogen means no water. Although we are far from certain about what conditions are necessary for life, most biologists believe that water is necessary. On the other hand, if the nuclear force were substantially weaker than what it actually is, then the complex atoms needed for biology could not hold together. As another example, if the relationship between the strengths of the gravitational force and the electromagnetic force were not close to what it is, then the cosmos would not harbor some stars that explode and spew out life-supporting chemical elements into space and other stars that form planets. Both kinds of stars seem necessary for the emergence of life. In sum, the strengths of the basic forces and certain other fundamental parameters in our universe appear to be fine-tuned to allow the existence of life.

The recognition of this fine-tuning led the British physicist Brandon Carter to articulate in 1968 what he called the anthropic principle, which states that the universe must have many of the parameters it does because we are here to observe it. Actually, the word "anthropic," stemming from the Greek word for "man," is a misnomer. If these fundamental parameters were much different from what they are, it is not only we human beings who would not exist. No life of any kind would exist.

If such conclusions are correct, the great question, of course, is *why* do these fundamental parameters happen to lie within the range needed for life? Does the universe care about life? Intelligent Design is one answer. Indeed, a number of theologians, philosophers, and even some scientists have used fine-tuning and the anthropic principle as evidence for the existence of God. For example, at the 2011 annual Christian Scholars' Conference at Pepperdine University, Francis Collins, a leading geneticist and director of the National Institutes of Health, said, "To get our universe, with all of its potential for complexities or any kind of potential for any kind of life form, everything has to be precisely defined on this knife edge of improbability . . . you have to see the hands of a Creator who set the parameters to be just so because the Creator was interested in something a little more complicated than random particles."

Intelligent Design is an answer to fine-tuning that does not appeal to most scientists. The multiverse offers another explanation. If there are zillions of different universes with different properties—for example, some with nuclear forces much stronger than in our universe and some with nuclear forces much weaker— then some of those universes will allow the emergence of life and some will not. Some of those universes will be dead, lifeless hulks of matter and energy, and some will permit the emergence of cells, plants and animals, minds. From the huge range of possible universes predicted by the theories, the fraction of universes with life is undoubtedly small. But that doesn't matter. We live in one of the universes that permits life because otherwise we wouldn't be here to ponder the question.

The explanation is similar to the explanation of why we happen to live on a planet that has so many nice things for our comfortable existence: oxygen, water, a temperature between the freezing and boiling points of water, and so on. Is this happy coincidence just good luck, or an act of providence, or what? No, it is simply that we could not live on planets without such properties. Many other planets exist that are not so hospitable to life, such as Uranus, where the temperature is −371 degrees Fahrenheit, or Venus, where the rain is sulfuric acid.

The multiverse idea offers an explanation to the fine-tuning conundrum that does not require the presence of a Designer. As Weinberg says: "Over many centuries science has weakened the hold of religion, not by disproving the existence of God, but by invalidating arguments for God based on what we observe in the natural world. The multiverse idea offers an explanation of why we find ourselves in a universe favorable to life that does not rely on the benevolence of a creator, and so if correct will leave still less support for religion."

Some physicists remain skeptical of the anthropic principle and the reliance on multiple universes to explain the values of the fundamental parameters of physics. Others, such as Weinberg and Guth, have reluctantly accepted the anthropic principle and the multiverse idea as together providing the best possible explanation for the observed facts.

If the multiverse idea is correct, then the historic

mission of physics to explain all the properties of our universe in terms of fundamental principles—to explain why the properties of our universe must *necessarily* be what they are—is futile, a beautiful philosophical dream that simply isn't true. Our universe is what it is simply because we are here. The situation can be likened to that of a group of intelligent fish who one day begin wondering why their world is completely filled with water. Many of the fish, the theorists, hope to prove that the cosmos necessarily has to be filled with water. For years, they put their minds to the task but can never quite seem to prove their assertion. Then a wizened group of fish postulates that maybe they are fooling themselves. Maybe, they suggest, there are many other worlds, some of them completely dry, some wet, and everything in between.

The most striking example of fine-tuning, and one that practically demands the multiverse to explain it, is the unexpected detection of what scientists call "dark energy." Little more than a decade ago, using robotic telescopes in Chile, Hawaii, Arizona, and outer space that can comb through nearly a million galaxies a night, astronomers discovered that the expansion of the universe is accelerating. As mentioned previously, it has been known since the late 1920s that the universe is expanding, a central aspect of the Big Bang model. Orthodox cosmological thought held that the expansion is slowing down. After all, gravity is an attractive force, which pulls masses closer together. So it was quite a surprise in 1998 when two teams of astronomers announced that some unknown force appeared to be jamming its foot down on the cosmic accelerator pedal. The expansion is speeding up. Galaxies are flying away from one another as if repelled by antigravity. Says Robert Kirshner, one of the team members, "This is not your father's universe." (In October 2011, members of both teams were awarded the Nobel Prize in Physics.)

Physicists call the energy associated with this unexpected cosmological force dark energy. No one knows what it is. Not only invisible, dark energy apparently hides out in empty space. Yet, based on our observations of the accelerating rate of expansion, dark energy comprises a whopping three-quarters of the total energy of the universe. Dark energy is the ultimate éminence grise. Dark energy is the invisible elephant in the room of science.

The amount of dark energy, or more precisely the amount of dark energy in every cubic centimeter of space, has been measured to be about one-hundred-millionth (10^{-8}) of an erg per cubic centimeter. (For comparison, a penny dropped from waist high hits the floor with an energy of about 300,000—that is, 3×10^5—ergs.) This may not seem like much, but it adds up in the vast volumes of outer space. Astronomers were able to determine this number by measuring the rate of expansion of the universe at different epochs. If the universe is accelerating, then its rate of expansion was slower in the past. From the amount of acceleration, astronomers can calculate the amount of dark energy.

Theoretical physicists have several hypotheses for the identity of dark energy. It may be the energy of ghostly subatomic particles that can briefly appear out of nothing before annihilating and slipping back into the vacuum. According to quantum physics, empty space is a pandemonium of subatomic particles, rushing about and then vanishing before they can be seen. Dark energy may also be associated with an hypothesized but as-yet-unobserved force field called the Higgs field, which is sometimes invoked to explain why certain kinds of matter have mass. Theoretical physicists ponder things that other people do not. [Note: A year after this essay was written, in the summer of 2012, physicists claimed to have observed the Higgs field. See "The Symmetrical Universe."] According to string theory, dark energy may be associated with the way in which extra dimensions of space—beyond the usual length, width, and breadth—get compressed down to sizes much smaller than atoms, so that we do not notice them.

These various hypotheses give a fantastically large range for the *theoretically possible* amounts of dark energy in a universe, from something like 10^{115} ergs per cubic centimeter to -10^{115} ergs per cubic centimeter. (A negative value for dark energy means that it acts to *decelerate* the universe, in contrast to what is observed.) Thus, in absolute magnitude, the amount of dark energy actually present in our universe is very, very small compared to what it could be. This fact alone is surprising. If the theoretically possible values for dark energy were marked out on a ruler stretching from here to the sun, the value of dark energy actually found in our universe (10^{-8} ergs per cubic centimeter) would be closer to the zero end than the width of an atom.

On one thing most physicists agree. If the amount of dark energy in our universe were only a little bit different than what it actually is, then life could never have emerged. A little larger, and the universe would have accelerated so rapidly that matter in the young universe could never have pulled itself together to form stars and hence complex atoms made in stars. And, going into negative values of dark energy, a little smaller and the universe would have decelerated so rapidly that it would have recollapsed before there was time to form even the simplest atoms.

Here we have a clear example of fine-tuning: out of all the possible amounts of dark energy that our universe might have, the actual amount lies in the tiny sliver of the range that allows life. There is little argument on this point. It does not depend on assumptions about whether we need liquid water for life or oxygen or particular biochemistries. It depends only on the requirement of atoms. As before, one is compelled to ask the question: Why does such fine-tuning occur? And the answer many physicists now believe: the multiverse. A vast number of universes may exist, with many different values of the amount of dark energy. Our particular universe is one of the universes with a small value, permitting the emergence of life. We are here, so our universe must be such a universe. We are an accident. From the cosmic lottery hat containing zillions of universes, we happened to draw a universe that allowed life. But then again, if we had not drawn such a ticket, we would not be here to ponder the odds.

The concept of the multiverse is compelling not only because it explains the problem of fine-tuning. As I mentioned earlier, the possibility of the multiverse is actually predicted by modern theories of physics. One such theory, called eternal inflation, is a revision of Guth's inflation theory developed by Paul Steinhardt, Alex Vilenkin, and Andrei Linde in the early and mid-1980s. In the inflation theory, the very rapid expansion of the infant universe is caused by an energy field, like dark energy, that is temporarily trapped in a condition that does not represent the lowest possible energy for the universe as a whole—like a marble sitting in a small dent on a table. The marble can stay there, but if it is jostled, it will roll out of the dent, roll across the table, and then fall to the floor (which represents the lowest possible energy level). In the theory of eternal inflation, the dark energy field has many different values at different points of space, analogous to lots of marbles sitting in lots of dents on the cosmic table. Each of these marbles is jostled by the random processes inherent in quantum mechanics, and some of the marbles will begin rolling across the table and onto the floor. Each marble starts a new Big Bang, essentially a new universe. Thus, the original, rapidly expanding universe spawns a multitude of new universes, in a never-ending process.

String theory, too, predicts the possibility of the multiverse. Originally conceived in the late 1960s as a theory of the strong nuclear force but soon enlarged far beyond that ambition, string theory postulates

that the smallest constituents of matter are not subatomic particles, like the electron, but extremely tiny
one-dimensional "strings" of energy. These elemental strings can vibrate at different frequencies, like the
strings of a violin, and the different modes of vibra-
tion correspond to different fundamental particles and
forces. String theories typically require seven dimensions of space in addition to the usual three, which
are compacted down to such small sizes that we never
experience them, like a three-dimensional garden hose
that appears as a one-dimensional line when seen from
a great distance. There are, in fact, a vast number of
ways that the extra dimensions in string theory can be
folded up, a little like the many ways that a piece of
paper can be folded up, and each of the different ways
corresponds to a different universe with different physical properties.

It was originally hoped that from a theory of these
strings, with very few additional parameters, physicists
would be able to explain all the forces and particles of
nature—all of reality would be a manifestation of the
vibrations of elemental strings. String theory would
then represent the ultimate realization of the Platonic
ideal of a fully explicable cosmos in terms of a few
fundamental principles. In the last few years, however,
physicists have discovered that string theory does not
predict a unique universe, but a vast number of pos-

sible universes with different properties. It has been estimated that the "string landscape" contains 10^{500} different possible universes. For all practical purposes, that number is infinite.

It is important to point out that neither eternal inflation nor string theory has anywhere near the experimental support of many previous theories in physics, such as general relativity or quantum electrodynamics. Eternal inflation or string theory, or both, could turn out to be wrong. However, some of the world's leading physicists have devoted their careers to the study of these two theories.

Back to the intelligent fish. The wizened old fish conjecture that there are many other worlds, some with dry land and some with water. Some of the fish grudgingly accept this explanation. Some feel relieved. Some feel like their lifelong ruminations have been pointless. And some remain deeply concerned. Because there is no way they can prove this conjecture. That uncertainty also disturbs many physicists who are adjusting to the idea of the multiverse. Not only must we accept that basic properties of our universe are accidental and uncalculable. In addition, we must believe in the existence of many other universes. But we have no conceivable way of observing these other universes and cannot prove their existence. Thus, to explain what we see in the world and in our mental deductions, we must believe in what we cannot prove.

Sound familiar? Theologians are accustomed to taking some beliefs on faith. Scientists are not. Such arguments, in fact, run hard against the long grain of science. All we can do is hope that the same theories that predict the multiverse also make other predictions that we can test here in our local universe. But the other universes themselves will almost certainly remain a conjecture.

"We had a lot more confidence in our intuition before the discovery of dark energy and the multiverse idea," says Guth. "There will still be a lot for us to understand, but we will miss out on the fun of figuring everything out from first principles." One wonders whether a twenty-five-year-old Alan Guth, entering science today, would choose theoretical physics.

The Temporary Universe

Last August my oldest daughter got married. The ceremony took place at a farm in the little town of Wells, in Maine, against the backdrop of rolling green meadows, a white wooden barn, and the sounds of a classical guitar. Each member of the wedding party stepped down a sloping hill toward the chuppah, while the guests sat in simple white chairs bordered by rows of sunflowers. The air was redolent with the smells of maples and grasses and other growing things. It was a marriage we had all hoped for. The two families had known each other with affection for years. Radiant in her white dress, a white dahlia in her hair, my daughter asked to hold my hand as we walked down the aisle.

It was a perfect picture of utter joy, and utter tragedy. Because I wanted my daughter back as she was at age ten, or twenty. As we moved together toward that lovely arch that would swallow us all, other scenes flashed through my mind: my daughter in first grade holding a starfish as big as herself, her smile missing a tooth; my daughter on the back of my bicycle as we rode to a river to drop stones in the water; my daughter telling me the day after she had her first period. Now she was thirty. I could see lines in her face.

I don't know why we long so for permanence, why the fleeting nature of things so disturbs. With futility, we cling to the old wallet long after it has fallen apart. We visit and revisit the old neighborhood where we grew up, searching for the remembered grove of trees and the little fence. We clutch our old photographs. In our churches and synagogues and mosques, we pray to the everlasting and eternal. Yet, in every nook and cranny, nature screams at the top of her lungs that nothing lasts, that it is all passing away. All that we see around us, including our own bodies, is shifting and evaporating and one day will be gone. Where are the one billion people who lived and breathed in the year 1800, only two short centuries ago?

The evidence seems overly clear. In the summer months, mayflies drop by the billions within twenty-four hours of birth. Drone ants perish in two weeks. Daylilies bloom and then wilt, leaving dead, papery stalks. Forests burn down, replenish themselves, then disappear again. Ancient stone temples and spires flake in the salty air, fracture and fragment, dwindle to spindly nubs, and eventually dissolve into nothing. Coastlines erode and crumble. Glaciers slowly but surely grind down the land. Once, the continents were joined. Once the air was ammonia and methane. Now it is oxygen and nitrogen. In the future, it will be something else. The sun is depleting its nuclear fuel. And just look at our own bodies. In the middle years and beyond, skin sags and cracks. Eyesight fades. Hearing diminishes. Bones shrink and turn brittle.

Just the other day, I had to retire my favorite shoes, a pair of copper-colored wingtips that I purchased thirty years ago to wear at a friend's graduation. For the first few years, all I had to do to keep the shoes looking spiffy was to polish them. Then the soles began to wear down. Every couple of years, I would take my wingtips to a small shoe repair shop I knew and have new soles installed. The shop was run by three generations of an Italian family. In the early years, the grandfather worked on my shoes. Then he died and his son took over the job. The resoling kept my shoes going another twenty years. My wife begged me to surrender. But I loved those shoes. They reminded me of me in my salad days. Eventually, the upper leather of the shoes became so thin that it cracked and split. I took the shoes back to the shop. The cobbler looked at them, shook his head, and smiled.

. . .

Physicists call it the second law of thermodynamics. It is also called the arrow of time. Oblivious to our human yearnings for permanence, the universe is relentlessly wearing down, falling apart, driving itself toward a condition of maximum disorder. It is a question of probabilities. You start from a situation of improbable order, like a deck of cards all arranged according to number and suit, or like a solar system with several planets orbiting nicely about a central star. Then you drop the deck of cards on the floor over and over again. You let other stars randomly whiz by your solar system, jostling it with their gravity. The cards become jumbled. The planets get picked off and go aimlessly wandering through space. Order has yielded to disorder. Repeated patterns to change. In the end, you cannot defeat the odds. You might beat the house for a while, but the universe has an infinite supply of time and can outlast any player.

Consider the world of living things. Why can't we live forever? The life cycles of amoebas and humans are, as everyone knows, controlled by the genes in each cell. While the raison d'être of the majority of genes is to pass on the instructions for how to build a new amoeba or human being, an important number of genes concern themselves with supervising cellular operations and replacing worn-out parts. Some of these genes must be copied thousands of times; others are constantly subjected to random chemical storms and electrically unbalanced atoms, called free radicals, that disrupt other atoms. Disrupted atoms, with their electrons misplaced, cannot properly pull and tug on nearby atoms to form the intended bonds and architectural forms. In short, with time the genes get degraded. They become forks with missing tines. They cannot quite do their job. Muscles, for example. With age, muscles slacken and grow loose, lose mass and strength, can barely support our weight as we toddle across the room. And why must we suffer such indignities? Because our muscles, like all living tissue, must be repaired from time to time due to normal wear and tear. These repairs are made by the mechano growth factor hormone, which in turn is regulated by the IGF1 gene. When that gene inevitably loses some tines . . . Muscle to flab. Vigor to decrepitude. Dust to dust.

In fact, most of our body cells are constantly being sloughed off, rebuilt, and replaced to postpone the inevitable. As one might imagine, the inner surface of the gut comes into contact with a lot of nasty stuff that damages tissue. To stay healthy, the cells that line this organ are constantly being renewed. Cells just below the intestine's surface divide every twelve to sixteen hours, and the whole intestine is refurbished every few days. I figure that by the time an unsuspecting person reaches the age of forty, the entire lining of her large intestine has been replaced several thousand times. Billions of cells have been shuffled each go-round. That makes trillions of cell divisions and whispered messages in the DNA to pass along to the next fellow in the chain. With such numbers, it would be nothing short of a miracle if no copying errors were made, no messages misheard, no foul-ups and instructions gone awry. Perhaps it would·be better just to remain sitting and wait for the end. No, thank you.

Yet despite all the evidence, we continue to strive for youth and immortality, we continue to cling to the old photographs, we continue to wish that our grown daughters were children again. Every civilization has sought the "elixir of life"—the magical potion that would grant youth and immortality. In China alone, the substance has one thousand names. It is known in Persia, in Tibet, in Iraq, in the aging nations of Europe. Some call it *Amrita*. Or *Aab-i-Hayat*. Or *Maha Ras*. *Mansarover*. *Chasma-i-Kausar*. *Soma Ras*. Dancing Water. Pool of Nectar. In the ancient Sumerian epic of *Gilgamesh,* one of the earliest known works of literature, the warrior king Gilgamesh goes on a difficult and dangerous journey in search of the secret of eternal life. At the end of Gilgamesh's journey, the flood god, Utnapishtim, suggests that the warrior king try out a taste of immortality by staying awake for six days and seven nights. Before Utnapishtim can finish the sentence, Gilgamesh has fallen asleep. In his old age, Qin Shi Huangdi, the first emperor of China, sent hundreds of minions in search of the elixir of life. When they returned empty-handed, his court doctors prescribed mercury pills to make him immortal, and he soon died of mercury poisoning. But he would eventually have died anyway.

We pay good money for toupees and tummy tucks, face-lifts and breast lifts, hair dyes, skin softeners, penile implants, laser surgeries, Botox treatments, injections for varicose veins. We ingest vitamins and pills and anti-aging potions and who knows what else. I recently did a Google search for "products to stay young" and got 37,200,000 hits.

But it is not only our physical bodies that we want frozen in time. Most of us struggle against change of all kinds, both big and small. Companies dread structural reorganization, even when it may be for the best, and have instituted whole departments and directives devoted to "change management" and the coddling of employees through tempestuous times. Stock markets plunge during periods of flux and uncertainty. "Better the devil you know than the devil you don't." Who among us clamors to replace the familiar and comfortable incandescent lightbulbs with the new, odd-looking, "energy-efficient" compact fluorescent lamps and light-emitting diodes? We resist throwing out our worn loafers, our thinning pullover sweaters, our child-hood baseball gloves. A plumber friend of mine will not replace his twenty-year-old water pump pliers, even though they have been banged up and worn down over the years. Outdated monarchies are preserved all over the world. In the Catholic Church, the law of priestly celibacy has remained essentially unchanged since the Council of Trent in 1563.

I have a photograph of the coast near Pacifica, Cali-fornia. Due to irreversible erosion, California has been losing its coastline at the rate of eight inches per year. Not much, you say. But it adds up over time. Fifty years ago, a young woman in Pacifica could build her house a safe thirty feet from the edge of the bluff overlooking the ocean, with a beautiful maritime view. Five years went by. Ten years. No cause for concern. The edge of the bluff was still twenty-three feet away. And she loved her house. She couldn't bear moving. Twenty years. Thirty. Forty. Now the bluff was only three feet away. Still she hoped that somehow, some way, the ero-sion would cease and she could remain in her home. She hoped that things would stay the same. In actual fact, she hoped for a repeal of the second law of ther-modynamics, although she may not have described her desires that way. In the photograph I am looking at, a dozen houses on the coast of Pacifica perch right on the very edge of the cliff, like fragile matchboxes, with their undersides hanging over the precipice. In some, awnings and porches have already slid over the side and into the sea.

Over its 4.5-billion-year history, our own planet has gone through continuous upheavals and change. The primitive Earth had no oxygen in its atmosphere. Due to its molten interior, our planet was much hotter than it is now, and volcanoes spewed forth in large numbers. Driven by heat flow from the core of the Earth, the terrestrial crust shifted and moved. Huge landmasses splintered and glided about on deep tectonic plates. Then plants and photosynthesis leaked oxygen into the atmosphere. At certain periods, the changing gases in the air caused the planet to cool, ice covered the Earth, entire oceans may have frozen. Today, the Earth continues to change. Something like ten billion tons of carbon are cycled through plants and the atmosphere every few years—first absorbed by plants from the air in the form of carbon dioxide, then converted into sugars by photosynthesis, then released again into soil or air when the plant dies or is eaten. Wait around a hundred million years or so, and carbon atoms are recycled through rocks, soil, and oceans as well as plants.

What about our sun and other stars? Shakespeare's *Julius Caesar* says to Cassius: "But I am constant as the northern star, / Of whose true-fix'd and resting quality / There is no fellow in the firmament." But Caesar was not up on modern astrophysics or the second law of thermodynamics. The North Star and all stars, including our sun, are consuming their nuclear fuel, after which they will fade into cold embers floating in space or, if massive enough, bow out in a final explosion. Our sun, for example, will last another five billion years before its fuel is spent. Then it will expand enormously into a red gaseous sphere, enveloping the Earth, go through a series of convulsions, and finally settle down to a cold ball made largely of carbon and oxygen. In past eons, new stars have replaced the dying stars by the action of gravity pulling together cosmic clouds of gas. But the universe has been expanding and thinning out since its Big Bang beginning, large concentrations of gas are gradually being disrupted, and in the future the density of gas will not be sufficient for new star formation. In addition, the lighter chemical elements that fuel most stars, such as hydrogen and helium, will have been used up in previous generations of stars. At some point in the future, new stars will cease being born. Slowly but surely, the stars of our universe are winking out. A day will come when the night sky will be totally black, and the day sky will be totally black as well. Solar systems will become planets orbiting dead stars. According to astrophysical calculations, in about a million billion years, plus or minus, even those dead solar systems will be disrupted from chance gravitational encounters with other stars. In about ten billion billion years, even galaxies will be disrupted, the cold spheres that were once stars flung out to coast solo through empty space.

Buddhists have long been aware of the evanescent nature of the world.

Anicca, or impermanence, they call it. In Buddhism, anicca is one of the three signs of existence, the others being *dukkha,* or suffering, and *anatta,* or non-selfhood. According to the Mahaparinibbana Sutta, when the Buddha passed away, the king deity Sakka uttered the following: "Impermanent are all component things. They arise and cease, that is their nature: They come into being and pass away." We should not "attach" to things in this world, say the Buddhists, because all things are temporary and will soon pass away. All suffering, say the Buddhists, arises from attachment.

If I could only detach from my daughter, perhaps I would feel better.

But even Buddhists believe in something akin to immortality. It is called Nirvana. A person reaches Nirvana after he or she has managed to leave behind all attachments and cravings, after countless trials and reincarnations, and finally achieved total enlightenment. The ultimate state of Nirvana is described by the Buddha as *amāravati,* meaning deathlessness. After a being has attained Nirvana, the reincarnations cease. Indeed, nearly every religion on Earth has celebrated the ideal of immortality. God is immortal. Our souls might be immortal.

To my mind, it is one of the profound contradictions of human existence that we long for immortality, indeed fervently believe that something must be unchanging and permanent, when all of the evidence in nature argues against us. I certainly have such a longing. Either I am delusional, or nature is incomplete. Either I am being emotional and vain in my wish for eternal life for myself and my daughter (and my wingtips), or there is some realm of immortality that exists outside nature.

If the first alternative is right, then I need to have a talk with myself and get over it. After all, there are other things I yearn for that are either not true or not good for my health. The human mind has a famous ability to create its own reality. If the second alternative is right, then it is nature that has been found wanting. Despite all the richness of the physical world—the majestic architecture of atoms, the rhythm of the tides, the luminescence of the galaxies—nature is missing something even more exquisite and grand: some immortal substance, which lies hidden from view. Such exquisite stuff could not be made from matter, because all matter is slave to the second law of thermodynamics. Perhaps this immortal thing that we wish for exists beyond time and space. Perhaps it is God. Perhaps it is what made the universe.

Of these two alternatives, I am inclined to the first. I cannot believe that nature could be so amiss. Although there is much that we do not understand about nature, the possibility that it is hiding a condition or substance so magnificent and utterly unlike everything else seems too preposterous for me to believe. So I am delusional. In my continual cravings for eternal youth and constancy, I am being sentimental. Perhaps with the proper training of my unruly mind and emotions, I could refrain from wanting things that cannot be. Perhaps I could accept the fact that in a few short years, my atoms will be scattered in wind and soil, my mind and thoughts gone, my pleasures and joys vanished, my "I-ness" dissolved in an infinite cavern of nothingness. But I cannot accept that fate even though I believe it to be true. I cannot force my mind to go to that dark place. "A man can do what he wants," said Schopenhauer, "but not want what he wants."

Suppose I ask a different kind of question: If against our wishes and hopes, we are stuck with mortality, does mortality grant a beauty and grandeur all its own? Even though we struggle and howl against the brief flash of our lives, might we find something majestic in that brevity? Could there be a preciousness and value to existence stemming from the very fact of its temporary duration? And I think of the night-blooming cereus, a plant that looks like a leathery weed most of the year. But for one night each summer its flower opens to reveal silky white petals, which encircle yellow lacelike threads, and another whole flower like a tiny sea anemone within the outer flower. By morning, the flower has shriveled. One night of the year, as delicate and fleeting as a life in the universe.

The Spiritual Universe

I

Ten years ago, I began attending monthly meetings of
a small group of scientists, actors, and playwrights in a
carpeted seminar room at the Massachusetts Institute
of Technology. Our raison d'être, broadly speaking,
has been an exploration of how science and art affect
each other. As the late afternoon sun drains from the
room, we discuss all manner of topics, ranging from
the history of scientific discovery to the nature of the
creative process to the way that an actor connects to an
audience to the latest theater in New York and Boston.
Our salon succeeds because we never have an agenda.
At the beginning of each session, one of us will begin
talking about some random idea, another person will
chime in or change the subject, and miraculously, after
twenty minutes, we find that we have zeroed in on a
question that everyone is passionate about.

What continues to astonish me is the frequency

with which religion slips into the room, unbidden but persistent. One member of our group, playwright and director Alan Brody, offers this explanation: "Theater has always been about religion. I am talking about the beliefs that we live by. And science is the religion of the twenty-first century." But if science is the religion of the twenty-first century, why do we still seriously discuss heaven and hell, life after death, and the manifestations of God? Physicist Alan Guth, another member of our salon, pioneered the inflation version of the Big Bang theory and has helped extend the scientific understanding of the infant universe back to a trillionth of a trillionth of a trillionth of a second after $t = 0$. A former member, biologist Nancy Hopkins, manipulates the DNA of organisms to study how genes control the development and growth of living creatures. Hasn't modern science now pushed God into such a tiny corner that He or She or It no longer has any room to operate—or perhaps has been rendered irrelevant altogether? Not according to surveys showing that more than three-quarters of Americans believe in miracles, eternal souls, and God. Despite the recent spate of books and pronouncements by prominent atheists, religion remains, along with science, one of the dominant forces that shape our civilization. Our little group of scientists and artists finds itself fascinated with these contrasting beliefs, fascinated with different ways of understanding the world. And fascinated by how science and religion can coexist in our minds.

As both a scientist and a humanist myself, I have struggled to understand different claims to knowledge. As part of that struggle, I have eventually come to a formulation of the kind of religious belief that would, in my view, be compatible with science. The first step in this journey is to state what I will call the central doctrine of science: All properties and events in the physical universe are governed by laws, and those laws are true at every time and place in the universe. Although scientists do not talk explicitly about this doctrine, and my doctoral thesis adviser never mentioned it once to his graduate students, the central doctrine is the invisible oxygen that most scientists breathe. We do not, of course, know all the fundamental laws at the present time. But most scientists believe that a complete set of such laws exists and, in principle, that it is discoverable by human beings, just as nineteenth-century explorers believed in the North Pole although no one had yet reached it.

An example of a scientific law is the conservation of energy: the total amount of energy in a closed system remains constant. The energy in an isolated container may change form, as when the chemical energy latent in a fresh match changes into the heat and light energy of a burning flame—but, according to the law of the conservation of energy, the total amount of energy does not change. At any moment in time, we regard our knowledge of the laws of science as provisional. And from era to era in the history of science, we have found that some of our "working" laws must be revised, such as the replacement of Newton's law of gravity (1687) by Einstein's deeper and more accurate law of gravity (1915). But such revisions are part of the process of science and do not undermine the central doctrine—that a complete set of laws does exist, and that those laws are inviolable. The title of a book by the Nobel Prize–winning physicist Steven Weinberg is *Dreams of a Final Theory.*

Next, a working definition of God. I would not pretend to know the nature of God, if God does indeed exist, but for the purposes of this discussion, and in agreement with almost all religions, I think we can safely say that God is understood to be a Being not restricted by the laws that govern matter and energy in the physical universe. In other words, God exists outside matter and energy. In most religions, this Being acts with purpose and will, sometimes violating existing physical law (that is, performing miracles), and has additional qualities such as intelligence, compassion, and omniscience.

Starting with these axioms, we can say that science and God are compatible as long as the latter is content to stand on the sidelines once the universe has begun. A God that intervenes after the cosmic pendulum has been set into motion, violating the physical laws, would clearly upend the central doctrine of science. Of course, the physical laws could have been created by God before the beginning of time. But once created, according to the central doctrine, the laws are immutable and cannot be violated from one moment to the next.

We can categorize religious beliefs according to the degree to which God acts in the world. At one extreme is *atheism:* God does not exist, period. Next comes *deism.* A prominent belief in the seventeenth and eighteenth centuries and partly motivated to incorporate the new science with theological thinking, deism holds that God created the universe but has not acted thereafter. Voltaire considered himself a deist. Moving in the direction of a more vigorous God, next comes *immanentism:* God created the universe and the physical laws and continues to act but only through repeated application of those fixed laws. (See, for example, *God's Activity in the World: The Contemporary Problem,* edited by Owen Thomas.) While immanentism differs philosophically from deism, it is functionally equivalent because God does not perform miracles in the world, and the central doctrine of science is upheld. One can argue that Einstein believed in an immanentist God. Finally comes what some theologians call *interventionism* (see, for example, Charles Hodge in *Systematic Theology*): From time to time, God can and does act to violate the laws.

Most religions, including Christianity, Judaism, Islam, and Hinduism, subscribe to an interventionist view of God. Following the discussion above, all of these religions, at least in their orthodox expressions, are incompatible with science. This is as far as one gets with a purely logical analysis. Except for a God who sits down after the universe begins, all other Gods conflict with the assumptions of science.

But the situation is more complex than this. The majority of religious nonscientists accept the value of science even though they do not appreciate or embrace the central doctrine. And some individual scientists believe in some physical events that cannot be analyzed by the methods of science or that even contradict science. In other words, some fraction of scientists reject the central doctrine of science. It turns out that a significant number of scientists living today are devoutly religious in the orthodox sense. A recent study by the Rice University sociologist Elaine Howard Ecklund, who interviewed nearly 1,700 scientists at elite American universities, found that 25 percent of her subjects believe in the existence of God.

Francis Collins, leader of the celebrated Human Genome Project and now director of the National Institutes of Health, recently told *Newsweek*, "I've not had a problem reconciling science and faith since I became a believer at age 27 . . . if you limit yourself to the kinds of questions that science can ask, you're leaving out some other things that I think are also pretty important, like why are we here and what's the meaning of life and is there a God? Those are not scientific questions. I simply would argue you need to be thoughtful when you're asking a question—is this a faith question or a science question? As long as one keeps that distinction clearly in mind, then I don't see a conflict." Ian Hutchinson, professor of nuclear science and engineering at MIT, told me: "The universe exists because of God's actions. What we call the 'laws of nature' are upheld by God, and they are our description of the normal way in which God orders the world. I do think miracles take place today and have taken place over history. I take the view that science is not all the reliable knowledge that exists. The evidence of the resurrection of Christ, for example, cannot be approached in a scientific way." Owen Gingerich, professor emeritus of astronomy and of the history of science at Harvard University, says: "I believe that our physical universe is somehow wrapped within a broader and deeper spiritual universe, in which miracles can occur. We would not be able to plan ahead or make decisions without a world that is largely law-like. The scientific picture of the world is an important one. But it does not apply to all events. Even in science we take a lot for granted. It's a matter of what you want to trust. Faith is about hope rather than proof."

Devoutly religious scientists, such as Collins, Hutchinson, and Gingerich, reconcile their belief in science with their belief in an interventionist God by adopting a worldview in which the autonomous laws of physics, biology, and chemistry govern the behavior of the physical universe *most* of the time and therefore warrant our serious study. However, on occasion, God intervenes and acts outside of these laws. The exceptional divine actions cannot be analyzed by the methods of science.

I will put my cards on the table. I am an atheist myself. I completely endorse the central doctrine of science. And I do not believe in the existence of a Being who lives beyond matter and energy, even if that Being refrains from entering the fray of the physical world. However, I certainly agree with Collins and Hutchinson and Gingerich that science is not the only avenue for arriving at knowledge, that there are interesting and vital questions beyond the reach of test tubes and equations. Obviously, vast territories of the arts concern inner experiences that cannot be analyzed by science. The humanities, such as history and philosophy, raise questions that do not have definite or unanimously accepted answers.

Finally, I believe there are things we take on faith, without physical proof and even sometimes without any methodology for proof. We cannot clearly show why the ending of a particular novel haunts us. We cannot prove under what conditions we would sacrifice our own life in order to save the life of our child. We cannot prove whether it is right or wrong to steal in order to feed our family, or even agree on a definition of "right" and "wrong." We cannot prove the meaning of our life, or whether life has any meaning at all. For these questions, we can gather evidence and debate, but in the end we cannot arrive at any system of analysis akin to the way in which a physicist decides how many seconds it will take a one-foot-long pendulum to make a complete swing. The previous questions are questions of aesthetics, morality, philosophy. These are questions for the arts and the humanities. These are also questions aligned with some of the intangible concerns of traditional religion.

As another example, I cannot prove that the central doctrine of science is true.

Years ago, when I was a graduate student in physics, I was introduced to the concept of the "well-posed problem": a question that can be stated with enough clarity and precision that it is guaranteed an answer. Scientists are always working on well-posed problems. It may take researchers decades or lifetimes to find the answer to a particular question, and science is constantly revising itself in accordance with new experimental data and new ideas, but I would argue that at any moment in time, every scientist is working on, or attempting to work on, a well-posed problem, a question with a definite answer. We scientists are taught from an early stage of our apprenticeship not to waste time on questions that do not have clear and definite answers.

But artists and humanists often don't care what the answer is because definite answers don't exist to all interesting and important questions. Ideas in a novel or emotion in a symphony are complicated with the intrinsic ambiguity of human nature. That is why we can never fully understand why the highly sensitive Raskolnikov brutally murdered the old pawnbroker in *Crime and Punishment,* whether Plato's ideal form of government could ever be realized in human society, whether we would be happier if we lived to be a thousand years old. For many artists and humanists, the question is more important than the answer. As the German poet Rainer Maria Rilke wrote a century ago,

"We should try to love the questions themselves, like locked rooms and like books that are written in a very foreign tongue." Then there are also the questions that have definite answers but which we cannot answer. The question of the existence of God may be such a question.

As human beings, don't we need questions without answers as well as questions with answers?

I imagine the conversation in the MIT seminar room, with the murmurings of students in the hall and the silent photographs of Einstein and Watson and Crick staring from the wood-paneled walls: *I agree with much of you've said, says Jerry, but we need to distinguish between physical reality and what's in our heads. Something like the resurrection of Christ is a physical event. It either happened or it didn't. But how do you know what is physical reality? says Debra. You sound like Bishop Berkeley, says Rebecca.*

Throughout history, philosophers, theologians, and scientists have proposed arguments for or against various religious beliefs. In recent years, especially with the advances in cosmology, biology, and evolutionary theory, a number of prominent scientists, in particular, have used science to counter arguments put forth to support the existence of God—Steven Weinberg, Sam

Harris, and Lawrence Krauss, to name a few. The most vocal of these thinkers and critics is the British evolutionary biologist and author Richard Dawkins.

In his widely read book *The God Delusion,* Dawkins employs modern biology, astronomy, evolutionary theory, and statistics to attack two common arguments for the existence of God: that only an intelligent and powerful Being could have designed the universe as we find it (the argument of Intelligent Design), and that only the action and will of God could explain our sense of morality and, in particular, our desire to help others in need. In brief, Dawkins shows that the various wondrous phenomena of the universe, including our own comfortable situation on Earth, could have arisen completely from the laws of nature and random processes, without the necessity of a supernatural and intelligent Designer. He further shows how our sense of morality and altruism could follow logically from the process of natural selection, applied to individual genes, without the need to invoke God.

In the case of our comfortable environment on Earth, for example, we and all life-forms on Earth are fortunate to have liquid water, which many biologists believe is necessary for life as we know it. Liquid water, in turn, requires that our planet be at a favorable distance from the sun, not so close that the resulting high temperature would exceed the boiling point of water and not so far away that the temperature would lie below the freezing point of water. Proponents of Intelligent Design have argued that such propitious conditions are evidence of a Designer who wanted life on Earth. Dawkins and other scientists have offered an alternative explanation. There are almost certainly billions upon billions of solar systems in our galaxy, with planets at many different distances from their central star. In most of those solar systems, none of the orbiting planets are at the right distance for liquid water, but in some, the distance is right. We live on such a planet. If we didn't, we wouldn't be here to ponder the situation. Although Dawkins is too smart to claim that he has disproved the existence of God, he does title a chapter of his book "Why There Almost Certainly Is No God."

As a scientist, I find Dawkins's efforts to rebut these two arguments for the existence of God—Intelligent Design and morality—completely convincing. However, as I think he would acknowledge, falsifying the arguments put forward to support a proposition does not falsify the proposition. Science can never know what created our universe. Even if tomorrow we observed another universe spawned from our universe, as could hypothetically happen in certain theories of cosmology, we could not know what created *our* universe. And as long as God does not intervene in the contemporary universe in such a way as to violate physical laws, science has no way of knowing whether God exists or not. The belief or disbelief in such a Being is therefore a matter of faith.

Richard Dawkins and others can expend as many calories as they wish arguing that God does not exist, but my guess is that they will convince few people who already have faith. Either such a person believes in a nonintervening God, in which case scientific arguments are irrelevant, or the person, like Dr. Collins and Professors Hutchinson and Gingerich, believes that God lives beyond the restrictions of matter and energy and scientific analysis. Dawkins's accomplishment, and I salute him for it, is to provoke more discussion of the topic and to help empower the expression of atheism.

What troubles me about Dawkins's pronouncements is his wholesale dismissal of religion and religious sensibility. In a speech at the Edinburgh International Science Festival in 1992, Dawkins said: "Faith is the great cop-out, the great excuse to evade the need to think and evaluate evidence. Faith is belief in spite of, even perhaps because of, the lack of evidence." And a month after September 11, 2001, Dawkins told the British newspaper *The Guardian:* "Many of us saw religion as harmless nonsense. Beliefs might lack all supporting evidence but, we thought, if people needed a crutch for consolation, where's the harm? September 11th changed all that." In such condescending comments as these, Dawkins seems to label people of faith as nonthinkers.

In my opinion, Dawkins has a narrow view of faith, and of people. I would be the first to challenge any belief that contradicts the findings of science. But, as I have said earlier, there are things we believe in that do not submit to the methods and reductions of science. Furthermore, faith and the passion for the transcendent that often goes with it have been the impulse for so many exquisite creations of humankind. Consider the verses of the *Gitanjali,* the *Messiah,* the mosque of the Alhambra, the paintings on the ceiling of the Sistine Chapel. Should we take to task Tagore and Handel and Sultan Yusuf and Michelangelo for not *thinking*? For believing in *nonsense,* to use Dawkins's language? Reaching beyond art to the world of public affairs, should we label as nonthinkers Abraham Lincoln, Mahatma Gandhi, and Nelson Mandela because of their religious beliefs, because of their faith in some things that cannot be proved? Can we not accept their value as powerful thinkers and doers even if we do not agree with all of their beliefs?

Faith, in its broadest sense, is about far more than belief in the existence of God or the disregard of scientific evidence. Faith is the willingness to give ourselves over, at times, to things we do not fully understand.

Faith is the belief in things larger than ourselves. Faith is the ability to honor stillness at some moments and at others to ride the passion and exuberance that is the artistic impulse, the flight of the imagination, the full engagement with this strange and shimmering world.

Scattered throughout Dawkins's writings are comments that religion has been a destructive force in human civilization. Certainly, human beings, in the name of religion, have sometimes caused great suffering and death to other human beings. But so has science, in the many weapons of destruction created by physicists, biologists, and chemists, especially in the twentieth century. Both science and religion can be employed for good and for ill. It is how they are used by human beings, by us, that matters. Human beings have sometimes been driven by religious passion to build schools and hospitals, to create poetry and music and sweeping temples, just as human beings have employed science to cure disease, to improve agriculture, to increase material comfort and the speed of communication.

For many years, a family of ospreys lived in a large nest near my summer home in Maine. Each season, I carefully observed their rituals and habits. In mid-April, the parents would arrive, having spent the winter in

South America, and lay eggs. In early June, the eggs hatched. The babies slowly grew, as the father brought fish back to the nest, and in early to mid August were large enough to make their first flight. My wife and I recorded all of these comings and goings with cameras and in a notebook. We wrote down the number of chicks each year, usually one or two but sometimes three. We noted when the chicks first began flapping their wings, usually a couple of weeks before flying from the nest. We memorized the different chirps the parents made for danger, for hunger, for the arrival of food. After several years of cataloguing such data, we felt that we knew these ospreys. We could predict the sounds the birds would make in different situations, their flight patterns, their behavior when a storm was brewing. Reading our "osprey journals" on a winter's night, we felt a sense of pride and satisfaction. We had carefully studied and documented a small part of the universe.

Then, one August afternoon, the two baby ospreys of that season took flight for the first time as I stood on the circular deck of my house watching the nest. All summer long, they had watched me on that deck as I watched them. To them, it must have looked like I was in my nest just as they were in theirs. On this particular afternoon, their maiden flight, they did a loop of my house and then headed straight at me with tremendous speed. My immediate impulse was to run for cover, since they could have ripped me apart with their powerful talons. But something held me to my ground. When they were within twenty feet of me, they suddenly veered upward and away. But before that dazzling and frightening vertical climb, for about half a second we made eye contact. Words cannot convey what was exchanged between us in that instant. It was a look of connectedness, of mutual respect, of recognition that we shared the same land. After they were gone, I found that I was shaking, and in tears. To this day, I do not understand what happened in that half second. But it was one of the most profound moments of my life.

II

In April 2012, as the magnolias were coming to full bloom, my birth state, Tennessee, adopted a new law protecting teachers who allow their students to challenge evolution, climate change, and other scientific theories. Of course, the questioning and testing of any body of knowledge is always a healthy activity. But thoughtful critics of the new law worry that it will tacitly give permission for schools to put creationism and evolution on equal footing and once more to confuse religion and science. All of which raises the perennial issue of the boundaries between science and religion. So what exactly are those boundaries? What are the different kinds of knowledge in science and in religion? And how do we come by those different kinds of knowledge?

These are not easy questions, and I have wrestled with them for much of my life. For many years, I have lived in the world of science as a physicist, and I have been trained in the methods and logic of science. I have also lived in the world of the arts and humanities as a novelist, and I understand that we have beliefs and experiences that exist beyond the reach of rational analysis.

Broadly speaking, there are two kinds of knowledge in science: the properties of physical things, and the laws that govern those physical things. The latter we call the laws of nature. For example, we know the size and mass of golf balls. We know the sounds made by nightingales. We know the colors of sunlight. In modern science, we arrive at these facts by measurement with scales and rulers and other devices outside of our bodies. In earlier centuries, we used our human sight and hearing and touch to fathom the world, but those senses vary from person to person and cannot be easily standardized. We might say that the color of sunlight appears yellowish to our eyes, with a bit of red, but a far more accurate and reliable method of determining the colors of sunlight is to pass that light through a prism and to use an electronic device to measure the amount of red light, the amount of yellow light, the amount of green light, and so on. As much as possible, science tries to determine the properties of things in ways that can be repeated over and over, always giving the same result.

The laws of nature are more abstract. They are mathematical rules about how matter and energy behave. In part I of this essay, I gave the conservation of energy as an example of a scientific law. Another is the law of gravity. Discovered by Isaac Newton in the seventeenth century, the law of gravity quantifies the force between objects based on their masses and distance apart. With knowledge of the law of gravity, for example, we can predict how long it will take our golf ball to hit the ground when dropped from a height of ten feet, or any other height, to an accuracy of many decimal places. We could also predict how long it would take the same golf ball to hit the dust when dropped on the moon, or on Mars. The central doctrine of science, as discussed previously, states that the laws of nature are the same everywhere in the universe.

The history of science has been a process of gradually discovering and revising the laws of nature. Often, we discover new laws by making guesses, inspired by our conceptions of simplicity or beauty or analogy with older laws. But then we must test those guesses against experiment. Some lovely guesses, such as the idea that the orbits of planets are circular, have been proven wrong by careful observation and experiment. As we develop new measuring devices, make better experiments, and reconceive our ideas of scientific principles, we constantly update and revise what we hold to be the laws of nature. The law of gravity, as formulated by Newton, is extremely accurate for most applications. But, as stated previously, it has been replaced by an even more accurate law discovered by Einstein a century ago. Einstein's law of gravity does not include quantum physics, and, at some point, it too will undoubtedly be replaced by a new law that does. At the present time, in the year 2014, we certainly do not know all the laws of nature, and it is a good bet that most of our current formulations of those laws will be revised in the future. Yet the great majority of scientists believe that a complete and final set of laws governing all physical phenomena exists, and that we are making continual progress toward discovery of those laws. That belief is part of the central doctrine of science.

Let me turn now to religion. In his landmark study of religion, *Varieties of Religious Experience* (1902), the great Harvard philosopher William James described religion in this way: "Were one to characterize religion in the broadest and most general terms possible, one might say that it consists of the belief that there is an unseen order, and that our supreme good lies in harmoniously adjusting ourselves thereto." As I will discuss shortly, the central role of "belief" in James's statement makes religion a fundamentally personal and subjective experience, which, with a few exceptions, distinguishes it from science.

I would suggest that there are two kinds of knowledge in religion: the transcendent experience, and the content of sacred religious books, such as the Old Testament of Judaism, the New Testament of Christianity, the Koran of Islam, and the Upanishads of Hinduism. Some religious leaders suggest that we should call religious knowledge "faith" or "intuitive knowledge" or "wisdom."

The transcendent experience—the immediate and vital personal experience of being connected to some unseen divine order—is beautifully described by a clergyman in James's book:

I remember the night, and almost the very spot on the hilltop, where my soul opened out, as it were, into the Infinite, and there was a rushing together of two worlds, the inner and the outer. It was deep calling unto deep—the deep that my own struggle had opened up within being answered by the unfathomable deep without, reaching beyond the stars. I stood alone with Him who had made me, and all the beauty of the world, and love, and sorrow, and even temptation. I did not seek Him, but felt the perfect union of my spirit with His . . . Since that time no discussion that I have heard of the proofs of God's existence has been able to shake my faith. Having once felt the presence of God's spirit, I have never lost it again for long. My most assuring evidence of his existence is deeply rooted in that hour of vision in the memory of that supreme experience.

The extremely personal and immediate nature of the transcendent experience described here is what gives it power and force. The clergyman who underwent the moment on the hilltop has no doubt of what he felt, and that remembered feeling represents a kind of truth, a knowledge of his own being and his felt connection to the cosmos. No other person can negate that personal experience. And no matter how the clergyman tries to analyze his experience with science or theology or references to sacred books, the experience is ultimately beyond analysis. The truth and power of it lies in the subjective experience itself. As James writes, "Our impulsive belief is here always what sets up the original body of truth, and our articulately verbalized

philosophy is but its showy translations into formulas."
The strong sense of the infinite, the belief in an unseen
order in the world, the feeling of being in the presence of something divine are all *personal*. Qualities of
this experience cannot be quantified or measured, like
readings on a voltmeter, and thus cannot transferred to
others. The qualities must be directly experienced by
the individual in unique moments.

Science also has something akin to the personal
experience, in the varied working styles of individual
scientists and in the feelings and passions of individual scientists for their work. Indeed, it is the personal
commitment of a scientist that keeps him or her up all
night in the lab or scribbling equations into the wee
hours of the morning. Such emotional and personal
involvement of scientists with their work is probably
essential to the scientific enterprise, as beautifully described in the book *Personal Knowledge,* by the highly
distinguished chemist Michael Polanyi. However, the
essence of science is the impersonal and the disembodied. Once the experiment has been completed or
the equation derived, no matter how emotionally attached the scientist who claims the discovery, no matter whether the scientist preferred working in the
morning or the afternoon, the results must be reproduced by other scientists in other conditions to gain
acceptance. Except for the field of psychology, science

concerns itself with the external world, outside our minds. Science indeed has a level of practice that is personal and human, but it also has an additional level of authentication, which is entirely impersonal and objective, and that additional level, existing outside of our minds, is what makes science science.

The sacred books of religion, another kind of religious knowledge, are sometimes treated as grand metaphors, sometimes as literal truth, sometimes as teachings of inspired human beings, sometimes as the direct words of God. Part of the content of the sacred books, such as the Ten Commandments or the advice of Krishna to Arjuna in the *Bhagavad Gita,* are prescriptions about how to live a moral life, or philosophies about meaning and value. Other content, such as the exodus of the Jews from Egypt around 1300 BC, or the Resurrection of Christ, deal with historical events. One can accept the statements about historical events without questioning or testing—in other words, without proof—in which case we might call that subjective knowledge, or perhaps belief. It would certainly not be scientific knowledge.

Science also engages in a few beliefs without proof: for example, belief in the central doctrine of science, as discussed in part I. There is no way that we can prove that the same laws of nature hold everywhere in the universe, since we cannot collect data from all parts of the universe. All of the data we have gathered from the farthest galaxies in the cosmos are consistent with a universal set of laws, but we cannot test every atom and molecule in the universe. Another tenet of faith in science is that the laws of nature are ultimately discoverable by us human beings. In Milton's *Paradise Lost,* when Adam asks the archangel Gabriel questions about celestial motions, Gabriel explains that studying the skies will reveal whether it is the Earth or the Heavens that rotate on their axes, but "the rest from Man or Angel the great Architect did wisely to conceal, and not divulge His secrets." In contrast to the admonitions of Gabriel, science believes that all knowledge about the physical world is within the province of human beings to discover. In science, no knowledge about the physical universe is off-limits or out of bounds.

Returning to the sacred books of religion as a possible source of knowledge, one can test historical statements against the same kind of evidence used by historians: authenticated documents and eyewitness reports written at the time, material relics that can be dated, the context of related events, plausibility, and so forth. Finally, if one considers the content of the sacred books to be metaphorical, then neither belief nor proof is needed. We are enlightened and uplifted by the narratives themselves, just we are by *The Tempest* of Shakespeare or the *Eroica* of Beethoven.

It is sometimes useful to distinguish between a physical universe and a spiritual universe, with the physical universe being the constellation of all physical matter and energy that scientists study, and the spiritual universe being the "unseen order" that James refers to, the territory of religion, the nonmaterial and eternal things that most humans have believed throughout the ages. The physical universe is subject to rational analysis and the methods of science. The spiritual universe is not. All of us have had experiences that are not subject to rational analysis. Besides religion, much of our art and our values and our personal relationships with other people spring from such experiences. I would argue, again, that the distinction between the spiritual and physical universes closely aligns with the axes of the personal and the impersonal. Events in the physical universe can be recorded with rulers and clocks and lie outside our bodies. Those measurements provide the evidence. Although many of us believe in a spiritual universe that hovers beyond our own personal being, the evidence of that universe is highly personal.

The physical and spiritual universes each have their own domains and their own limitations. The question of the age of planet Earth, for example, falls squarely in the domain of science, since there are reliable tests we can perform, such as using the rate of disintegration of radioactive rocks, to determine a definitive answer. Such questions as "What is the nature of love?" or "Is it moral to kill another person in time of war?" or "Does God exist?" lie outside the bounds of science but fall well within the realm of religion. I am impatient with people who, like Richard Dawkins, try to disprove the existence of God with scientific arguments. Science can never prove or disprove the existence of God, because God, as understood by most religions, is not subject to rational analysis. I am equally impatient with people who make statements about the physical universe that violate physical evidence and the known laws of nature. Within the domain of the physical universe, science cannot hold sway on some days but not on others. Knowingly or not, we all depend on the consistent operation of the laws of nature in the physical universe day after day—for example, when we board an airplane, allow ourselves to be lofted thousands of feet in the air, and hope to land safely at the other end. Or when we stand in line to receive a vaccination against the next season's influenza.

Some people believe that there is no distinction between the spiritual and physical universes, no distinction between the inner and the outer, between the subjective and the objective, between the miraculous and the rational. I need such distinctions to make sense of my spiritual and scientific lives. For me, there is room for both a spiritual universe and a physical universe, just as there is room for both religion and science. Each universe has its own power. Each has its own beauty, and mystery. A Presbyterian minister recently said to me that science and religion share a sense of wonder. I agree.

The Symmetrical Universe

One night after I had joined the staff of the Harvard College Observatory in Cambridge, I went up to the roof of the building and peered out of the telescope installed there in 1847. It was my first experience with a large telescope. (I was a theorist.) And there in the eyepiece, looking as big as a dinner plate, floated the planet Saturn, encircled by its delicate rings. The beauty dazzled me—the planet round as any roundness could be, the orbiting rings as symmetrical as any circles could be. How could nature create such perfection, without human meddling or mind? And why do we humans find the roundness of planets and rings so appealing?

There are, of course, many other symmetries in nature. Snowflakes exhibit perfect six-sided symmetry: each fragile branch is identical to the others. Small hailstones are round. Starfish have five equally spaced arms, each like the rest. Jellyfish divide into four identical sectors. The yellow iris has three petals and perfect three-sided symmetry: rotate the flower by one-third of a circle and it comes back to itself. Cut an apple in two, and you will find that its five seeds are arranged in a pentagonal pattern. The two wings of a butterfly. One could go on. Such pervasive symmetries could not be accidents.

I was reminded of cosmic symmetry last July, when scientists announced the discovery of the long-sought "Higgs boson," a subatomic particle hypothesized fifty years ago and whose existence is necessary for modern theories in physics. Although not mentioned in popular reports, one of the main functions served by the Higgs is to allow physicists to construct theories that embody profound symmetry.

Although each Higgs particle is far smaller than an atom, it takes a colossal machine to find one. That is because other subatomic particles—protons—must be accelerated up to nearly the speed of light and then crashed into each other to produce a Higgs. The only particle accelerator in the world with enough size and energy for that feat is the Large Hadron Collider near Geneva, Switzerland, built by the European Organization for Nuclear Research (CERN). The Large Hadron Collider winds around for 17 miles in a tunnel 545 feet below the ground on the Swiss-French border. It turns out that the Higgs particle is a shy little fellow. It takes roughly a trillion collisions between protons to coax one Higgs into existence, and, once created, the particle hangs around for less than a billionth of a trillionth of a second before changing into other subatomic particles. Clearly, a particle with such a fleeting acquaintance cannot be spotted directly. Rather, its existence is inferred by observing the other particles that it morphs into.

On July 4, 2012, two independent teams of scientists, each with about three thousand physicists, announced that they had discovered the tracks of a few Higgs particles in the debris from trillions of proton-proton collisions. "We're reaching into the fabric of the universe at a level we've never done before," said Joe Incandela, professor of physics at the University of California at Santa Barbara and leader of one of the two international teams. "We're on the frontier now, on the edge of a new exploration. This could be the only part of the story that's left, or we could open a whole new realm of discovery."

The "story" that Incandela mentions is called the Standard Model of physics, which gives a full accounting of most of the fundamental forces and particles of nature. (The four fundamental forces, as understood by modern physicists, are the gravitational force; the electromagnetic force; the "strong force," which traps the subatomic particles at the centers of atoms; and the "weak force," which is responsible for certain kinds of radioactive decay of atoms.)

In 1964, when the Standard Model was not yet even on the drawing board, Peter Higgs of the University of Edinburgh and other physicists theorized the existence of a new type of energy that would bestow mass on certain subatomic particles and leave others, like the photons of light, without mass. (Physicists worry about such things as why some particles have mass and others do not.) What later came to be called the Higgs particle was a consequence of the mass-giving energy. Then, in 1967, American physicist Steven Weinberg and Pakistani physicist Abdus Salam independently proposed a major piece of the Standard Model, a theory that united the weak force and the electromagnetic force within a common framework now called the "electroweak force."

In their proposed synthesis of the forces of nature, Weinberg and Salam were guided by an almost religious devotion to symmetry. And that devotion required the Higgs particle. Here's why. At its deepest level, the meaning of a symmetry is that you can make some change in a system and everything still looks the same, like swapping two arms of a starfish or rotating a snowflake by 60 degrees. The essence of the electroweak theory is the postulate that nature is symmetrical with respect to the particles that convey the weak and electromagnetic forces, known as Ws and Zs and photons, respectively. That is, you can exchange some of these particles with the others, and the fundamental forces act in the same way. In terms of the unified electroweak force, these particles are equivalent.

The only problem with Weinberg and Salam's proposal is that we know that photons and Ws and Zs are not identical, not like the branches of a snowflake. In particular, they have very different masses, so that we can easily distinguish one from the other. But by incorporating the Higgs energy into the theory, the difference in masses can be attributed to different amounts of friction with the Higgs rather than to a lack of underlying equivalence of the particles. The underlying symmetry is still there, and the Weinberg-Salam theory is built upon that symmetry. More important, the theory's predictions have been confirmed by experiment. The theory correctly predicted the properties of the W and Z particles as well as new kinds of interactions between those particles. In 1979, the two scientists and a third, Sheldon Glashow, were jointly awarded the Nobel Prize for their work on this theory. The only remaining question was whether the postulated Higgs particle, upon which everything depended, actually existed. As of early 2013, almost all physicists agree that the Higgs particle has at last been found, in the experiments at CERN. If it had continued to elude discovery, then not only the Standard Model would have been called into question but also the physicists' faith in the deep symmetry upon which that theory is based.

Some physicists believe that nature is even more symmetrical than we have yet discovered, that at high enough energy all four of the fundamental forces become essentially identical, with the same strength. Weinberg, arguably the greatest apostle of symmetry in the history of science, believes that symmetry principles are more fundamental than matter and energy and force. In his 1992 book, *Dreams of a Final Theory*, he writes:

> Symmetry principles have moved to a new level of importance in this [twentieth] century . . . there are symmetry principles that dictate the very existence of all the known forces of nature . . . We believe that, if we ask why the world is the way it is and then ask why that answer is the way it is, at the end of this chain of explanations we shall find a few simple principles of compelling beauty.

It is not difficult to understand why scientists like Weinberg are attracted to symmetry. For one thing, symmetry is associated with beautiful mathematics. As a simple example, consider the equation for a circle of radius R: $x^2 + y^2 = R^2$. (If you don't remember your high school math, no worries. Just view the equation as a picture.) Because a circle appears unchanged when rotated by any angle, this equation also embodies a rotational symmetry. If the x and y axes are rotated to make new axes, w and z, like rotating the north and east compass headings of a map, the equation for the same circle in the new coordinate system is $w^2 + z^2 = R^2$, exactly the same form as the original. What could be more lovely? The symmetry embedded in Weinberg and Salam's equations for the electroweak theory is similar, only a bit more involved. All theoretical scientists—those who work principally with mathematics—delight in the beauty of mathematics.

Scientists, especially physicists and more especially twentieth-century physicists, have been attracted to symmetry for another very practical reason: theories with symmetry have usually turned out to conform to nature—that is, to make predictions that agree with experiment. Relativity, Einstein's theory of time, and quantum chromodynamics, the theory of the strong force, are examples. Both embody strong symmetries, and both have been borne out by experimental test. Symmetry also reduces complexity. A physical system with right-left symmetry, for example, needs only half as many parameters to specify it as a system with no symmetry. In the symmetrical case, specify the right side and the left side is known. Theoretical scientists, whether they be physicists or chemists or biologists, prefer economy in their theories of nature, prefer theories with the minimum possible number of parts and parameters and principles. The fewer parameters and principles needed to specify a system, the greater the understanding.

As a mundane but still surprising example of symmetry, consider the symmetry of one-dollar bills. All one-dollar bills are equivalent. Any one-dollar bill can be exchanged for any other one-dollar bill, and it will have the same buying power. The system of commerce is unchanged. As one of the many consequences of this symmetry, various kinds of goods can be compared to one another once reduced to a value in dollars, all of which are equivalent. The replacement of the barter system of trade with a monetary system, somewhere around 3000 BC, represented a huge simplification in the exchanges between human beings and also an understanding of what is at stake in purchases and sales. In this case, the symmetry was artificially imposed by human beings.

The deep question is: Why does nature embody so much symmetry? We do not know the full answer to this question. However, we have some partial answers. Symmetry leads to economy, and nature, like human beings, seems to prefer economy. If we think of nature as a vast ongoing experiment, constantly trying out different possibilities of design, then those designs that cost the least energy or that require the fewest different parts to come together at the right time will take precedence, just as the principle of natural selection says that organisms with the best ability to survive will dominate over time. On the other hand, as far as we know, the symmetries in the electroweak theory and relativity and chromodynamics did not evolve from ongoing experiments with different designs. Rather, they were apparently built in at the origin of the universe, by whatever processes and principles determined the fundamental laws of physics (see "The Accidental Universe"). As I will discuss below, some symmetries in nature derive from mathematical theorems and truths. And it is hard to imagine any universe without the order of mathematics and logic.

One physical principle that governs nature over and over is the "energy principle": nature evolves to minimize energy. If you place some marbles on a flat table, after some time has passed you will find most of the marbles on the floor. That's because a marble on the floor is closer to the center of the Earth and has lower gravitational energy than on the table. Snowflakes have six-sided symmetry because of the angles that the two hydrogen atoms make with the oxygen atom in each water molecule. Those angles minimize the total electrical energy of the water molecule. Any other angles would produce greater energy. Large bodies, like the planet Saturn, are round because a spherical shape minimizes the total gravitational energy. A mathematical theorem says that a sphere is the particular geometrical shape that has the least surface area for a given volume. Many objects in nature, like hailstones and soap bubbles, have greater electrical energy the greater the surface area. Thus hailstones and soap bubble minimize their energy by having spherical shapes.

A beautiful illustration of some of the ideas above is the beehive. Each cell of a honeycomb is a nearly perfect hexagon, a space with six identical and equally spaced walls. Isn't that surprising? Wouldn't it be more plausible to find cells of all kinds of shapes and sizes, fitted together in a haphazard manner? It is a mathematical truth that there are only three geometrical figures with *equal sides* that can fit together on a flat surface without leaving gaps: equilateral triangles, squares, and hexagons. Any gaps between cells would be wasted space. Gaps would defeat the principle of economy. Now you might ask why the sides of a cell in a beehive need to be equal in length. It is possible that each cell could have a random shape and unequal sides and the next cell could then be custom made to fit to that cell, without gaps. And so on, one cell after another, each fit to the one before it. But this method of constructing a honeycomb would require that the worker bees work sequentially, one at a time, first making one cell, then fitting the next cell to that, and so on. This procedure would be a waste of time for the bees. Each insect would have to wait in line for the guy in front to finish his cell. If you've ever seen bees building a beehive (or watched a video of bees on YouTube), they don't wait for one another. They work simultaneously. So the bees need to have a game plan in advance, knowing that all the cells will fit together automatically. Only equilateral triangles, squares, and hexagons will do.

But why hexagons? Here unfolds another fascinating story. More than two thousand years ago, in 36 BC, the Roman scholar Marcus Terentius Varro conjectured that the hexagonal grid is the unique geometrical shape that divides a surface into equal cells with the smallest total perimeter. And the smallest total perimeter, or smallest total length of sides, means the smallest amount of wax needed by the bees to construct their honeycomb. For every ounce of wax, a bee must consume about eight ounces of honey. That's a lot of work, requiring visits to thousands of flowers and much flapping of wings. The hexagon minimizes the effort and expense of energy. But Varro had made only a conjecture. Astoundingly, Varro's conjecture, known by mathematicians as the Honeycomb Conjecture, was proven only recently, in 1999, by the American mathematician Thomas Hales. The bees knew it was true all along.

There's more to the bee story. Bees are related to the question of why flowers have so much symmetry. Bees need flowers for their food and for making wax, and flowers need bees for pollination. Experiments published in 2004 by researchers at the Freie Universität in Berlin and the CNRS Université Paul-Sabatier in Toulouse show that bees are more attracted to flowers with symmetry. And why are bees attracted to flowers with more symmetry? The same researchers propose that symmetrical stimuli from the flowers are more easily processed by the visual system in the bee brain—that is, they require less neurological apparatus. Again, the principle of economy at work.

But why are *we* attracted to symmetry? Why do we human beings delight in seeing perfectly round planets through the lens of a telescope and six-sided snowflakes on a cold winter day? The answer must be partly psychological. I would claim that symmetry represents order, and we crave order in this strange universe we find ourselves in. The search for symmetry, and the emotional pleasure we derive when we find it, must help us make sense of the world around us, just as we find satisfaction in the repetition of the seasons and the reliability of friendships. Symmetry is also economy. Symmetry is simplicity. Symmetry is elegance.

And however we define the mysterious quality that we call beauty, we associate symmetry with beauty. Both Darwin and Freud have argued that our sense of beauty and the appeal of beauty originated with the imperative for sexual reproduction and the association of beauty with a vibrant mate. As Darwin wrote in the *Descent of Man,*

> A sense of beauty has been declared to be peculiar to man. But when we behold male birds elaborately displaying their plumes and splendid colors before the females, while other birds not so decorated make no such display, it is impossible to doubt that the females admire the beauty of their male partners. As women everywhere deck themselves with these plumes, the beauty of such ornaments cannot be disputed.

Clearly, human-made art and architecture abound with symmetry. The Taj Mahal has a central dome and arch, two identical side domes, and four identical towers, symmetrically placed. Leading to the building is a rectangular pool with equally spaced cypress trees on both sides of the pool and symmetrical gardens beyond. The Octagon on Roosevelt Island in New York, designed by Alexander Jackson Davis, is shaped like a you-know-what. Leonardo da Vinci's famous "Vitruvian Man" depicts a male figure with two identical sets of outstretched and equally spaced arms and legs, one set inscribed within a circle and one within a square. The mosaic floor of the great cathedral at Cologne has a stunning set of nested circles filled with symmetrically placed flowers. A widely reproduced image of Lakshmi shows the Hindu goddess sitting in the center of a circular flower with two identical arms raised upward and holding identical yellow flowers, two more identical arms lowered and releasing flower petals, and two identical elephants on each side of her pouring water from identical jugs. However, if one looks at the image closely, it will be seen that there is a slight departure from perfect symmetry. Lakshmi has a red scarf draped over her left shoulder but not her right.

In fact, in human-made art, especially in painting, it seems that a slight bit of asymmetry is desirable and found to achieve a higher aesthetic satisfaction. Ernst

Gombrich, one of the leading (Western) art historians of the twentieth century, believes that although human beings have a deep psychological attraction to order, perfect order in art is uninteresting. "However we analyse the difference between the regular and the irregular," he writes, "we must ultimately be able to account for the most basic fact of aesthetic experience, the fact that delight lies somewhere between boredom and confusion. If monotony makes it difficult to attend, a surfeit of novelty will overload the system and cause us to give up." My wife, a painter trained in the tradition of the Boston School dating back to the early 1900s, always tells me that a well-designed painting should have some off-center and asymmetrical accent. Of course, asymmetry can be defined only relative to symmetry, and vice versa. Asymmetric elements in paintings or buildings are most effective when superimposed against a background of symmetry. Perhaps nature is being the painter when she occasionally violates complete symmetry with irregular coastlines and the amorphous shapes of clouds.

We should also point out that the association of symmetry with beauty in art is partly cultural. In some non-Western cultures, asymmetries can be as lovely as symmetries. No obvious symmetry can be found in the Great Wall of China, for example. Instead, it was built to conform to its natural terrain. The wall wanders and curves with the shape of the land, and its towers are irregularly spaced. It blends with its surroundings. The Chinese sense of beauty is, in some ways, more subtle, ambiguous, and less articulated than that of the West. For example, the world of the living is considered to be in symmetrical balance to the world of the dead. On the other hand, even in the Chinese artistic tradition, we find some obvious symmetries, such as the couplet in classical Chinese poetry, where verb is aligned with verb, noun with noun, rhyme with rhyme.

I find myself now looking at an old photograph taken in 1949. I am a baby, held in my mother's lap. Standing directly behind her is *her* mother, and on her right and left are her two grandmothers, my great-grandmothers—a symmetrical arrangement of five people. I study the faces, looking for more symmetry, or lack of it. Of course, there are the familiar symmetries of the human head. I look more closely. One of my great-grandmothers, called Oma, has a mouth that droops slightly on the left side, breaking the symmetry of her face. I associate that droop with the sadness of losing her husband only a few years into her marriage. If I look even more closely at the photograph, I can see a blemish on her right cheek, possibly an age spot, also breaking the symmetry. But these slight asymmetries announce themselves only against the background of symmetry.

In the end, it is easier to explain why bees construct honeycombs shaped like perfect hexagons than why human beings place identical towers on the sides of the Taj Mahal or the two grandmothers on equal sides of the mother. The first is a result of economy and mathematics, the second of psychology and aesthetics. Perhaps in asking why the pervasive symmetries in nature are found appealing to the human mind and imitated in our human-made constructions, we are making an erroneous distinction between our minds and the remainder of nature. Perhaps we are all the same stuff. After all, our minds are made of the same atoms and molecules as everything else in nature. The neurons in our brains obey the same physical laws as planets and snowflakes. Most important, our brains developed out of nature, out of hundreds of millions of years of sensory response to sunlight and sound and tactile connection to the world around our bodies. And the architecture of our brains was born from the same trial and error, the same energy principles, the same pure mathematics that happen in flowers and jellyfish and Higgs particles. Viewed in this way, our human aesthetic is *necessarily* the aesthetic of nature. Viewed in this way, it is nonsensical to ask why we find nature beautiful. Beauty and symmetry and minimum principles are not qualities we ascribe to the cosmos and then marvel at in their perfection. They are simply what is, just like the particular arrangement of atoms that make up our minds. We are not observers on the outside looking in. We are on the inside too.

The Gargantuan Universe

My most vivid encounter with the vastness of nature occurred years ago in the Aegean Sea. My wife and I had chartered a sailboat for a two-week holiday in the Greek Islands. After setting out from Piraeus, we headed south and hugged the coast, which we held three or four miles to port. In the thick summer air, the distant shore appeared as a hazy beige ribbon, not entirely solid but a reassuring line of reference. With binoculars, we could just make out the glinting of houses, fragments of buildings.

Then we passed the tip of Cape Sounion and turned west toward Hydra. Within a couple of hours, both the land and all other boats disappeared. Looking around in a full circle, all we could see was water, extending out and out in all directions until it joined with the sky. I felt insignificant, misplaced, a tiny odd trinket in a cavern of ocean and air.

Naturalists, biologists, philosophers, painters, and poets have labored to express the qualities of things in this strange world that we find ourselves in. Some things are prickly, others are smooth. Some are round, some jagged. Luminescent, or dim. Mauve-colored. Pitter-patter in rhythm. Of all of these aspects of things, none seems more immediate and vital than *size*. Large versus small. Consciously and unconsciously, we routinely measure our physical size against the dimensions of other people, animals, trees, oceans, mountains. As brainy as we think ourselves, our bodily size, our bigness, our simple volume and bulk are the first carrying cards we present to the world. I would hazard a guess that somewhere in our fathoming of the cosmos, we must keep a mental inventory of plain size and scale, going from atoms to microbes to us humans to oceans to planets to stars. And some of the most impressive additions to that inventory have occurred at the high end. Simply put, the cosmos has gotten larger and larger. At each new level of distance and scale, we have had to contend with a different conception of the world that we live in.

The prize for exploring the greatest distance in space goes to a man named Garth Illingworth, who works in a ten-by-fifteen-foot office at the University of California at Santa Cruz. Professor Illingworth studies galaxies so distant that their light has traveled though space for more than thirteen billion years to get from there to here. You can hardly turn around in his office. It is cramped with several tables and chairs, bookshelves, computers, scattered papers, copies of *Nature* magazine, and a small refrigerator and microwave to fortify himself for research that can extend into the wee hours of the morning.

Like most professional astronomers these days, Illingworth does not look directly through a telescope. He gets his images by remote control—in his case, quite remote. The telescope he uses is the Hubble Space Telescope, which orbits the Earth once every ninety-seven minutes, high above the distorting effects of the Earth's atmosphere. Hubble takes digital photographs of galaxies and radios these images to other orbiting satellites, which relay them to a network of earthbound antennae; these, in turn, send their signals to the Goddard Space Flight Center in Greenbelt, Maryland. From there the data are uploaded to a special website that Illingworth can access from a computer in his office.

The most distant galaxy Illingworth has seen so far goes by the name of UDFj-39546284, documented in early 2011. This galaxy is about 100,000,000,000,000, 000,000,000 miles away from Earth, give or take. It appears as a faint red blob against the speckled night of the distant universe—red because the light has been stretched to longer and longer wavelengths as it makes its lonely journey through space for billions of years. The actual color of the galaxy is blue, the color of young, hot stars, and it is twenty times smaller than our galaxy, the Milky Way. UDFj-39546284 was one of the first galaxies to form in the universe.

"That little red dot is hellishly far away," Professor Illingworth told me recently. At the age of sixty-five, Illingworth is a friendly bear of a man, with a ruddy complexion, a thick mane of strawberry-blond hair, wire-rimmed glasses, and a broad smile. "I sometimes think to myself: What would it be like to be out there, looking around?"

One measure of the progress of human civilization is the increasing scale of our maps. A clay tablet dating from the twenty-fifth century BC and found in what is now the city of Kirkuk, Iraq, depicts a river valley between two hills, with a plot of land labeled as 354 *iku* (about 30 acres) in size. In the earliest recorded cosmologies, such as the Babylonian Enuma Elish from around 1500 BC, the oceans, the continents, and the heavens were considered limited in size, but there were no scientific estimates of those sizes. The early Greeks, including Homer, viewed the Earth as a circular plate with the ocean enveloping it and Greece at the center, but there was no understanding of scale. In the early sixth century BC, the Greek philosopher Anaximander, considered the first mapmaker, and his student Anaximenes proposed that the stars were attached to a giant crystalline sphere. But again there was no figure of its size.

The first large object ever accurately measured was the Earth, accomplished in the third century BC by Eratosthenes, a geographer who administered the great library in Alexandria. From travelers, Eratosthenes had heard the intriguing report that at noon on June 21, in the town of Syene, due south of Alexandria, the sun cast no shadow at the bottom of a deep well. Evidently, the sun is directly overhead at that time and that place. (Before the invention of clocks, "noon" could be defined at each place as the moment when the sun was highest in the sky, whether that was exactly vertical or not.) Eratosthenes knew that the sun was not overhead at noon in Alexandria. In fact, it was tipped 7.2 degrees from the vertical, or about one-fiftieth of a circle—something he could determine by measuring the length of the shadow cast by a stick in the ground. That the sun could be overhead in one place and not at another was due to the curve of the Earth. Eratosthenes then reasoned that if he knew the distance from Alexandria to Syene, the full circumference of the Earth must be about fifty times that distance. Traders passing through Alexandria told him that camels could make the trip to Syene in about fifty days, and it was known that a camel could cover one hundred stadia (approximately 11.3 miles) in a day. So the ancient geographer estimated that Syene and Alexandria were about 570 miles apart. Consequently, the complete circumference of the Earth he figured to be about 50 × 570 miles, or 28,500 miles. This number was within 15 percent of the modern measurement, an amazing feat considering the imprecision of using camels as odometers.

As ingenious as they were, the ancient Greeks were not able to calculate the size of our solar system. That discovery had to wait nearly two thousand years for the invention of the telescope. In 1672, the French astronomer Jean Richer determined the distance to Mars by measuring how much the position of the planet shifted against the background of stars from two different observation points on Earth. The two points were Paris (of course) and Cayenne, French Guiana. Using the distance to Mars, astronomers were able to compute the distance from the Earth to the sun, approximately 100 million miles.

A few years later, Isaac Newton managed to estimate the distance to the nearest stars. (Only someone as accomplished as Newton could have been the first to perform such a calculation and have it go almost unnoticed among his other achievements.) If one assumes that the stars are similar objects to our sun, equal in intrinsic luminosity, Newton asked, how far away would our sun have to be in order to appear as faint as nearby stars? Writing his computations in a spidery script, with a quill dipped in the ink of oak galls, Newton correctly concluded that the nearest stars are about one hundred thousand times the distance from Earth to the sun, or roughly ten trillion miles away. Newton's calculation is contained in a short section of his *Principia,* titled simply "On the Distance of the Stars."

Newton's estimate of the distance to nearby stars was larger than any distance imagined before in human history. Even today, nothing in our experience allows us to relate to it. The fastest most of us have traveled is about five hundred miles per hour, the speed of a jet airplane. If we set out for the nearest star beyond our solar system at that speed, it would take about five million years to reach our destination. If we traveled in the fastest rocket ship ever manufactured on Earth, the trip would take one hundred thousand years, at least a thousand human life spans.

But even these lengths are dwarfed by the distances measured in the early twentieth century by Henrietta Leavitt, an astronomer at the Harvard College Observatory. In 1912, she devised a completely new method to determine the distances to faraway stars. Certain stars, called Cepheid variables, were known to oscillate in brightness. Leavitt discovered that the cycle times of such stars are closely related to their intrinsic luminosities. More luminous stars have longer cycle times. Measure the cycle time of such a star and you know its intrinsic luminosity. Then, by comparing its intrinsic luminosity to how bright it appears in the sky, you can infer its distance, just as you could gauge the distance to an approaching car in the night if you knew the wattage of its headlights. Cepheid variables are scattered throughout the cosmos. They serve as cosmic distance signs in the highway of space.

Using Leavitt's method of measuring great distances, astronomers in the next few years were able to determine the size of our galaxy, the Milky Way, which is a giant congregation of about 200 billion stars. To express such mind-boggling sizes and distances, twentieth-century astronomers adopted a new unit of distance called the light-year, the distance that light travels in a year—about six trillion miles. In these units, the nearest stars are several light-years away. The diameter of the Milky Way has been measured to be about one hundred thousand light-years. In other words, it takes a ray of light one hundred thousand years to travel from one side of the Milky Way to the other.

There are galaxies beyond our own. They have names like Andromeda (one of the nearest), Sculptor, Messier 87, Malin 1, IC 1101. The average distance between galaxies, again determined by Leavitt's method, is about twenty galactic diameters, or two million light-years. To a giant cosmic being, leisurely strolling through the universe and not limited by distance or time, galaxies would appear as illuminated mansions scattered about the dark countryside of space. As far as we know, galaxies are the largest objects in the cosmos. If we sorted the long inventory of material objects in nature by size, we would start with subatomic particles like electrons and end up with galaxies.

Over the last century, astronomers have been able to probe deeper and deeper into space, looking out to distances of hundreds of millions of light-years and further. The question naturally arises as to whether the physical universe could be unending in size. That is, as we build bigger and bigger telescopes, sensitive to fainter and fainter light, will we continue to see objects farther and farther away—like the third emperor of the Ming Dynasty, Yongle, who surveyed his new palace in the Forbidden City and walked from room to room to room, never reaching the end?

Here we must take into account a curious relationship between distance and time. Because light travels at a fast but still not infinite speed, 186,000 miles per second, when we look at a distant object in outer space, a significant amount of time has passed between the emission of the light and the reception at our end. The image we see is what the object looked like when it first emitted that light. If we look at an object 186,000 miles away, we see it as it appeared one second earlier; at 1,860,000 miles away, we see it as it appeared ten seconds earlier; and so on. For extremely distant objects, we see them as they were millions and billions of years in the past.

Now, the second curiosity. Since the late 1920s, we have known that the universe is expanding, and thinning out and cooling as it does so. By measuring the rate of expansion, we can make good estimates of the moment in the past when the expansion began—the Big Bang—which was about 13.7 billion years ago, a time when no planets or stars or galaxies existed and the entire universe consisted of a fantastically dense nugget of pure energy. No matter how big our telescopes, we cannot see beyond the distance light has traveled since the Big Bang beginning of the universe. Farther than that, and there simply hasn't been enough time since the birth of the universe for light to get from there to here. This giant sphere, the maximum distance we can see, is only the *observable* universe. (Each day, the observable universe gets a bit larger.) But the universe could extend far beyond that.

In his office at Santa Cruz, Garth Illingworth and his colleagues have mapped out and measured the cosmos to the edge of the observable universe. They have reached out almost as far as the laws of physics allow. All that exists in the knowable universe—oceans and sky, planets and stars, pulsars, quasars, dark matter, distant galaxies and clusters of galaxies, great clouds of star-forming gas—has been gathered within the cosmic sensorium gauged and observed by human beings.

"Every once in a while," says Professor Illingworth, "I think: By God, we are studying things that we can never physically touch. We sit on this miserable little planet in a midsized galaxy, and we can characterize most of the universe. It is astonishing to me, the immensity of the situation, and how to relate to it in terms we can understand."

The idea of Mother Nature has been represented in every culture on Earth. But to what extent is the new universe, vastly larger than anything conceived in the past, part of *nature*? One wonders how connected Illingworth feels to this fantastically large cosmic terrain, to the galaxies and stars so distant that their images have taken billions of years to reach our eyes. Are the "little red dots" on Illingworth's space maps part of the same landscape that Wordsworth and Thoreau described, part of the same visceral ethos as mountains and trees, part of the same cycle of birth and demise that orders our lives, part of our personal physical and emotional conception of the world that we live in? Or are such things instead digitized abstractions, silent and untouchable, akin to us only in their (hypothesized) makeup of atoms and molecules? And to what extent are we human beings, living on a small planet orbiting one star among billions of stars, part of that same nature?

Once, the heavenly bodies were considered divine, made of entirely different stuff than objects on Earth. Aristotle argued that all terrestrial substances were composed of four elements: earth, fire, water, and air. He reserved a fifth element, the "ether," for the heavenly bodies, which he considered immortal, perfect, and indestructible. It wasn't until the birth of modern science, in the seventeenth century, that we began to understand the similarity of heaven and earth. In 1610, using his new telescope, Galileo noted that the sun had dark patches and blemishes, destroying the belief in the perfection of the heavenly bodies. In 1686, Isaac Newton proposed a universal law of gravity, applying equally to the fall of an apple from a tree and to the orbits of planets around the sun. Newton went further, suggesting that all of the laws of nature apply to phenomena in the heavens as well as on Earth. In later centuries, scientists used our terrestrial understanding of chemistry and physics to estimate how long the sun could continue to shine before depleting its resources of energy; to determine the chemical composition of stars; to map out the formation of galaxies.

Yet even after Galileo and Newton, the question remained: Were living things somehow different from rocks and water and stars? Did animate and inanimate matter differ in some fundamental way? The "vitalists" claimed that animate matter had some special essence, an intangible spirit or soul, while the "mechanists" argued that living things were elaborate machines and obeyed precisely the same laws of physics and chemistry as inanimate material. In the late nineteenth century, two German physiologists, Adolf Eugen Fick and Max Rubner, independently began testing the mechanistic hypothesis by painstakingly tabulating the energies required for muscle contraction, body heat, and other physical activities and comparing these energies against the chemical energy stored in food. Each gram of fat, carbohydrate, and protein had its energy equivalent. By the end of the nineteenth century, Rubner concluded that the energy used by a living creature exactly equaled the energy consumed in its food. Living creatures were to be viewed as complex arrangements of biological pulleys and levers, electrical currents, and chemical energies. Our bodies are made of the same atoms and molecules as stones, water, and air.

And yet many had a lingering feeling that human beings were somehow separated from the rest of nature. Such a view is nowhere better illustrated than in the painting *Tallulah Falls* (1841) by American painter George Cooke, an artist associated with the Hudson River School. While this group of artists celebrated nature, they also believed that human beings were set apart from the natural world. Cooke's painting depicts tiny human figures standing on a little promontory above a deep canyon. The people are dwarfed by tree-covered mountains, massive rocky ledges, and a raging waterfall pouring down to the canyon below. Not only insignificant in size compared with their surroundings, the human beings are mere witnesses to a scene they are not part of and could never be a part of. Just a few years earlier, Ralph Waldo Emerson published his famous essay "Nature," an appreciation of the natural world that nonetheless held human beings separate from nature, at the very least in the moral and spiritual domain: "Man is fallen; nature is erect."

Today, with various "back to nature" movements attempting to resist dislocations brought about by modern technology, and with a worldwide awareness of global warming and other environmental problems, many people feel a new sympathy with the natural world on this planet. But the gargantuan cosmos beyond remains remote. We might understand at some intellectual level that those tiny points of light in the night sky are similar to our sun, made of the same atoms as our bodies, and that the cavern of outer space extends from our galaxy of stars to other galaxies of stars, to distances that would take light rays millions and billions of years to traverse. We might understand these discoveries in intellectual terms, but they are baffling abstractions, even disturbing, like the notion that each of us once was the size of a dot, without mind or thought. Science has vastly expanded the scale of our cosmos, but our emotional reality is still limited by what we can touch with our bodies in the time span of our lives. Bishop Berkeley, the eighteenth-century Irish philosopher, argued that the entire cosmos is a construct of our minds, that there is no material reality outside our thoughts. As a scientist, I cannot accept that belief. At the emotional and psychological level, however, I can have some sympathy with Berkeley's views. Modern science has revealed a world as far removed from our bodies as colors are from the blind.

Very recent scientific findings have added yet another dimension to the question of our place in the cosmos. For the first time in the history of science, we are able to begin making plausible estimates of the rate of occurrence of life in the universe. In March 2009, NASA launched a spacecraft, *Kepler,* with the mission to search for planets orbiting in the "habitable zone" of other stars. The habitable zone is the region in which the temperature is not so cold as to freeze water and not so hot as to boil it. For many reasons, biologists and chemists believe that liquid water is required for the emergence of life, even if that life is very different from life on Earth. Dozens of candidates for such planets have been found, and we can make a rough preliminary estimate that something like 3 percent of all stars are accompanied by a life-sustaining planet. The totality of living matter on Earth—not only humans but all animals, plants, bacteria, and pond scum—makes up about 0.00000001 percent of the mass of the planet. Combining this figure with the results from the *Kepler* spacecraft, and assuming that all life-sustaining planets do indeed have life, we can conclude that the fraction of stuff in the visible universe that exists in living form is something like 0.000000000000001 percent, or one-millionth of one-billionth of 1 percent. If some cosmic intelligence created the universe, life would seem to be only an afterthought. And if life emerged by random processes, vast amounts of lifeless material were needed for each particle of life. Such findings cannot help but bear upon the question of our significance in the universe.

Decades ago, when I was sailing with my wife in the Aegean Sea, in the midst of unending water and sky, I had a slight inkling of infinity. It was a sensation I had not experienced before, accompanied by feelings of awe, transcendence, fear, sublimity, disorientation, alienation, and disbelief. I set a course for 255 degrees, trusting in my compass—a tiny disk of painted numbers with a sliver of rotating metal—and hoped for the best. In a few hours, as if by magic, a pale ochre smidgeon of land appeared dead ahead, a thing that drew closer and closer, a place with houses and beds and other human beings.

The Lawful Universe

When I joined the faculty of MIT, I was given a dual appointment in science and in the humanities. Some days I would teach a physics class in the morning and a fiction-writing class in the afternoon. In the mornings, the universe was reduced to the irrefutable and almost obsessively regular motion of pendulums on strings, oscillations of springs, ripples of electromagnetic waves traveling through space—all described to high accuracy by equations I could write down with white chalk on the board. I talked to my students about a world of pure logic, pure reason, pure cause and effect. It was a world in which, except at the quantum level of the atom, the future was completely determined by the past and the inexorable churning of the laws of nature. No one objected. In the afternoons, I would walk across the courtyard to the humanities building (Building 14 in MIT parlance) and talk to my students about the messy nature of human affairs. The dimly lit alleys of the mind. Greed, jealousy, love thwarted, happiness, revenge, complex and ambiguous motives for action. Students who wrote stories with self-consistent characters, characters whose movements could be predicted and who always acted with rationality and reason, were roundly rebuked for having created nothing more than lifeless hunks of pulp. Real people are unpredictable, I said. A character who always acts rationally is a fraud. Any character you fully understand is as good as dead. Is that clear?

But aren't we made of the same particles and electricity whose trajectories and flows can be charted out and computed to mind-numbing accuracy? I would hazard the guess that not many of us *Homo sapiens* would leap at the chance to have our thoughts and behavior reduced to neat lines and mathematical symbols on the blackboard. In almost everything else, we strive for logic and pattern and quantification. We admire principles and laws. We embrace reasons and causes—some of the time. At other times, we value spontaneity, unpredictability, unlimited and unconstrained behavior, complete personal freedom. On the subject of rules and patterns, I think we are absolutely schizophrenic. We are drawn to the symmetry of a snowflake, and we are also drawn to the amorphous shape of a cloud floating high in the sky. We appreciate the regular features of animals of a pure breed, and we are also fascinated by hybrids and mongrels that do not fit into any classification scheme. We honor those people who have lived upright and sensible lives, and we also esteem the mavericks who have broken the mold. In some perplexing and ill-understood manner, we human beings with our oversized craniums seem to have a fondness both for the predictable and the unpredictable, the rational and the irrational, regularity and irregularity. Yes, we are certainly a difficult mess of self-contradictions.

Back to the marks on the blackboard. Let us look for a moment at the extreme of our rational side. Physics. Masses and forces. Action and reaction. Over the centuries, physicists have discovered rules for the fundamental forces in the cosmos, such as gravity, electricity and magnetism, and the nuclear force that keeps the particles at the centers of atoms from flying apart. No physical phenomena we have ever observed do not fall within the grasp of these rules. Some of the rules are still being revised, and we certainly do not yet have a complete understanding of the physical universe, but the present version of the rules can accurately predict the results of experiments with fundamental particles and forces to many decimal places. And the rules are quantitative. For example, Coulomb's law for electricity says that the strength of the electrical force between two charged particles decreases by a factor of four when the distance between them doubles (in mathematical form, $F = q_1 q_2 / r^2$). It is a rule that has been arrived at by many experiments and the logic of electromagnetic theory, and it is able to predict how electrically charged particles will affect each other anywhere in the universe.

As another example, which you can test for yourself, drop a weight to the floor from a height of 4 feet and time the duration of its fall. You should get about 0.5 seconds. From a height of 8 feet, you should get about 0.7 seconds. From a height of 16 feet, about 1 second. Repeat from several more heights and you will discover the rule that the time exactly doubles with every quadrupling of the height, a rule found by Galileo in the seventeenth century. With this rule, you can now predict the time to fall from any height. You have witnessed, firsthand, the lawfulness of nature.

We call these rules the "laws of nature." It is an interesting terminology. The concept of a law goes back at least four thousand years, to the ancient Assyrians and their Code of Ur-Nammu. Those first laws were, of course, rules for behavior in human society and were quantifiable only in the number of shekels of silver owed or quarts of salt poured into the mouth for each specified infraction. For example: "If a man proceeded by force and deflowered the virgin slavewoman of another man, that man must pay five shekels of silver." Our ancient ancestors knew also about rules of geometry. The Babylonians understood that the ratio of a circle's circumference to its diameter is a universal number (which we denote by π). Any circle drawn anywhere obeyed this relation. The Babylonians knew the Pythagorean theorem, relating the sides of a right triangle. These were the antecedents of "laws."

The idea of "nature," as I have discussed in "The Gargantuan Universe," is complex and multilayered in meaning. Roughly speaking, we can think of nature as the totality of the physical universe, animate and inanimate. So the "laws of nature" are universal rules for the physical universe. We do not simply invent laws of nature, the way the Assyrians decided that it was socially unacceptable to deflower virgins by force. We discover laws by a combination of theorizing and experiment. In the end, the experimental test of a provisional law is its crucial rite of passage. The discovery and articulation of the laws of nature has become one of the great achievements of human civilization: The Great Wall of China. *King Lear*. The Taj Majal. The *Mona Lisa*. Relativity.

I would suggest that even for the laws of nature—that most exacting expression of rationality in the world and of our faith in the power of reason—we are buffeted by conflicting and ambivalent desires. A case in point is the recent reported discovery of the "Higgs boson." (See "The Symmetrical Universe" for a more complete discussion of Higgs.) Theorized in 1964 by physicist Peter Higgs of the University of Edinburgh, the Higgs boson is a type of subatomic particle whose existence is required by the so-called Standard Model of physics, which is our most up-to-date version of the laws of nature. According to the theory, the Higgs boson and the energy associated with it provide the mechanism by which most elementary particles are endowed with mass. Without the Higgs, there would be no atoms, no planets, no stars. And if the Higgs boson doesn't exist, it's back to the drawing board for some of our laws of nature.

In July 2012, when two teams of physicists announced that they had discovered a new particle that might be the long-sought Higgs boson, many physicists jumped up and down with joy. But not all. Maria Spiropulu, a professor of physics at the California Institute of Technology and a member of one of the discovery teams, said to *The New York Times:* "I personally do not want [the new particle] to be Standard Model anything—I don't want it to be simple or symmetric or as predicted. I want us all to have been dealt a complex hand that will send me and all of us in a good loop for a long time." Professor Spiropulu was not alone in her wayward thoughts. We like order, but we also like surprises. We like the predictable. But we also like the unpredictable. Every once in a while, we demand a fly in the ointment.

One of my favorite accounts of ancient scientific thinking is the long poem *De rerum natura,* or *On the Nature of Things,* written by the Roman poet and philosopher Lucretius in about 50 BC. Cicero read *De rerum natura,* as did many other Romans. In his poem, Lucretius elucidates a theory of atoms, which were theorized to be tiny, indestructible units of matter out of which everything else was made. (The idea of atoms went back several centuries earlier, to Democritus and Epicurus.) The elemental atoms were thought to come in a variety of sizes, shapes, and textures and thus were able to explain the different properties of matter. But Lucretius had more on his agenda than the explanation of matter. For Lucretius, atoms were a defense against the two greatest fears of human beings at the time (and possibly still today): fear of the capricious meddling of the gods in human affairs, and fear of everlasting punishment of the soul after a questionable life on Earth. Atoms, because of their materiality and indestructibility, countered both fears. Since everything was held to be made of atoms, and atoms could not be created from nothing, the gods could not make things appear out of thin air, could not act on Earth without due process of cause and effect. From Lucretius:

> This terror of mind [fear of the gods and fear of death] and this gloom must be dispelled, not by the sun's rays or the bright shafts of day, but by the aspect and law of nature. The first principle of our study we will derive from this: that no thing is ever by divine power produced from nothing. For assuredly, a dread holds all mortals thus in bond, because they behold many things happening in heaven and earth whose causes they can by no means see, and they think them to be done by divine power. For which reasons, when we shall perceive that nothing can be created from nothing, then we shall at once more correctly understand from that principle what we are seeking, both the source from which each thing can be made and the manner in which everything is done without the working of the gods.

A bit later in the poem, Lucretius says that the mind and the spirit are also made of atoms, so that upon death, just as "mist and smoke disperse into the air, believe that the spirit also is spread abroad and passes away far more quickly and is more speedily dissolved into its atoms." Thus, there is no immortal soul left after death. We are made of nothing but atoms, and when we die, the atoms disperse in the wind. "Therefore, death is nothing to us."

For Lucretius, atoms were part of the laws of nature, and the laws of nature freed human beings from the quirks and the power of the gods. Although Lucretius certainly believed in the existence of divine beings, he argued that the laws of nature operated outside their influence. By contrast, most religious people today would argue that the laws of nature fall fully within the power of God. God, being the creator of all things, created the laws of nature, and God can violate the laws of nature whenever God chooses to do so. As Owen Gingerich, professor emeritus of astronomy and of the history of science at Harvard University, says: "I believe that our physical universe is somehow wrapped within a broader and deeper spiritual universe, in which miracles can occur. We would not be able to plan ahead or make decisions without a world that is largely law-like. The scientific picture of the world is an important one. But it does not apply to all events."

The change of religious belief from the polytheism of the ancient Romans and Egyptians and Babylonians to the monotheism of Judaism, Christianity, and Islam must have played a role in the understanding of the laws of nature. The laws of nature are the polar opposite of capriciousness and whim. With many gods, each with his or her own personality and whims, there is much more room for unpredictable divine behavior and consequent surprises on Earth than with a single god. With a single god, we human beings need to understand only a single divine consciousness. Little wonder that Lucretius, a believer in the pantheon of divinities of Roman mythology, was so eager to preach a philosophy that would liberate human beings from the intervention of the gods.

One of the earliest quantitative laws of nature was Archimedes's principle about the buoyant force of water, stated in his book *On Floating Bodies* in 250 BC: "Any body wholly or partially immersed in a fluid experiences an upward force equal to the weight of the fluid displaced." In his treatise "On Burning Mirrors and Lenses," published in AD 984, the Persian physicist Ibn Sahl gave an accurate and quantitative law for the angle by which light is deflected in traveling from one medium to another.

The figure of Isaac Newton must surely be considered a landmark in the emerging concept of a lawful universe. Newton's law for gravity was not only one of the first mathematical expressions of a fundamental force underlying the motions of bodies. It was also the first proposal that a rule for the behavior of material bodies on Earth should apply in the heavens as well—that is, the first real understanding of the universality of a law of nature. Part of Newton's brilliance was in recognizing that the same force that caused an apple to fall from a tree also caused the moon to orbit the Earth. Yet even Newton, master logician and reductionist, believed that the laws of nature were not sufficient to explain everything in the physical world. After many pages of calculations, Newton comes to the end of *The Principia,* the *General Scholium,* and confesses that the synchronized performance of moons and planets could never be explained by "mere mechanical causes" but requires "the counsel and dominion of an intelligent and powerful Being." In particular, Newton believed that friction would slowly degrade the motions of planets over time without the active intervention of God: "Motion is much more apt to be lost than got and always on the decay . . . Blind fate could never make all the planets move one and the same way in orbs concentric . . . Some inconsiderable irregularities [in planetary orbits] . . . will [be] apt to increase till the system wants a reformation" by God. So, even for Newton, while the laws of nature autonomously governed the physical universe most of the time, every now and then God intervened for a bit of editorial work.

A hundred years later, the need for the intervention of God in the orbits of planets and, by extension, in the operation of all of the laws of nature was eliminated by the French mathematician and scientist Pierre-

Simon Laplace, sometimes called the French Newton. Laplace, who did not win friends when he announced that he was the best mathematician in France, carefully calculated the orbits of planets, taking into account their mutual gravitational jostling of each other as well as their response to the sun. His conclusion was that the solar system was completely stable all by itself, under the gravity described by Newton's laws. Gravitational friction would not disrupt the system of planets. Divine intervention was unnecessary. According to the nineteenth-century British mathematician Augustus De Morgan, the story was passed around Paris that when Laplace presented his great book on celestial mechanics to Napoleon, the emperor (who enjoyed asking embarrassing questions) slyly mentioned that he'd been told the book made no mention of God. To which Laplace replied, "Je n'avais pas besoin de cette hypothèse-là" ("I have no need for that assumption").

In the twentieth century, with the discovery of laws for how time and space contract and expand with motion and gravity (relativity), laws for the microscopic behavior of subatomic particles (quantum mechanics), and laws for the forces that hold atomic nuclei together (quantum chromodynamics), physicists have codified their understanding and faith in the laws of nature. So strong is that faith that scientists are profoundly disturbed when it appears that one of the established laws has been violated. The conservation of energy is such a law. This law was discovered in the mid-nineteenth century as the result of independent experiments by the German physician Julius Robert Mayer and by James Prescott Joule, the British scion of a wealthy brewing family, who furnished his laboratory from inherited money. As discussed in "The Spiritual Universe," the law says that although energy can change from one form to another, the total amount of energy in an isolated container remains constant. Over the last couple of centuries, we have discovered how to quantify the amount of energy in motion, in heat, in gravity, and in many other phenomena, and the total in a closed system doesn't change. If you put a bomb that has eleven units of chemical energy in an impenetrable box and detonate the bomb, a split second later the chemical energy of the bomb will have transformed into the light and the motion and the heat of the flying debris, but the total amount of energy will be still be eleven units. The conservation of energy is one of the sacred cows of science. Since the mid-nineteenth century, it has been deeply embedded in all the other laws of science.

In 1914, physicists discovered what appeared to be a violation of the law of conservation of energy. Certain kinds of radioactive atoms were found to spit out subatomic particles called "beta particles." The energy of such an atom before and after the emission could be measured. According to the law of conservation of energy, the energy of the beta particle should equal the difference in atomic energies before and after, just as the difference in bank balances at two different times should be equaled by the total expenditure of money during that period. Against these expectations, the energy of the beta particle was found to vary all over the place, sometimes being one number and sometimes another. Some physicists repeated the measurements and got the same upsetting results. Others argued that the beta particles were indeed emitted with the correct energy but lost some of it in random collisions with other atoms before being measured. A small group of distinguished physicists reluctantly proposed that perhaps the law of the conservation of energy was valid only in an average sense but not for each event in each atom.

In December 1930, just before a major scientific conference in Europe, the Austrian prodigy Wolfgang Pauli wrote a letter to his colleagues about the troubling dilemma of beta emission. His letter begins: "Dear Radioactive Ladies and Gentlemen . . . I have hit upon a desperate remedy to save the . . . law of conservation of energy." Pauli then goes on to propose that when a radioactive atom emits a beta particle, it also emits another kind of particle, previously unknown and now called a neutrino, and the sum of the energies of the neutrino and the beta particle correctly equals the difference in atomic bank balances. In other words, some of the energy expenditures had been accounted for but others had not. The proposal of a new kind of fundamental particle in physics is not taken lightly. "I agree that my remedy could seem incredible because one should have seen those [neutrinos] much earlier if they really exist. But only the one who dares can win . . ." Pauli ends his letter with an apology to his colleagues. He will have to miss the conference in Tubingen because he is "indispensable" at a ball in Zurich.

Physicists weaned on the conservation of energy jumped at Pauli's invisible neutrino and even began building it into new theories of radioactive atoms. The neutrino remained only a hopeful dream until 1956, when American physicists Clyde Cowan and Frederick Reines detected it at the Savannah River nuclear reactor in South Carolina. And the law of the conservation of energy remained supreme.

The laws of nature help us create sanity in this strange cosmos we find ourselves in. The laws of nature protect us from the vagaries of the gods. The laws of nature satisfy a deep emotional need for order and reason and control.

. . .

Then there is the contrary in us. In their excellent book *Wonders and the Order of Nature,* historians of science Lorraine Daston and Katharine Park detail human-kind's fascination with wonders and oddities. Things that don't fit. Surprises and peculiarities. Marco Polo enthuses over finding completely black lions in the Indian kingdom of Quilon. James of Vitry reports on the strange "midnight sun" of Iceland, men with tails in Britain, women with huge goiters in the Burgun-dian Alps. Other travelers excitedly record gourds with little lamblike animals inside, beasts with the faces of humans and the tails of scorpions, unicorns, men with heads as hairy as dogs, petrifying lakes, colored moun-tains, plants that produce hallucinations, waters that cure disease, the powers of planets in juxtaposition, people who vomit worms, virgin births, powders that sexually arouse. And on and on. In his essay "Of Mir-acles" (1748), the Scottish philosopher David Hume writes that "the passion of surprise and wonder, aris-ing from miracles, being an agreeable emotion, gives a sensible tendency towards the belief of those events from which it is derived." More recently, the French philosopher Michel Foucault has written, "Curiosity pleases me. It evokes . . . a readiness to find strange and singular what surrounds us; a certain relentlessness to break up our familiarities." Daston and Park make the case that the attraction to marvels and miracles is more prevalent in the ignorant and has diminished over the centuries. I suggest that if we enlarge the category of miracles to include surprises and observations that do not fit within conventional thinking or known explanations, such unruly attractions still exist today, and in quite a few educated and civilized people. Consider Professor Spiropulu of Caltech. Or poet Wallace Stevens, who wrote: "It is the mundo of the imagination in which the imaginative man delights and not the gaunt world of reason." Or the recent Pew survey showing that two-thirds of Americans believe in supernatural events.

Certainly, Professor Spiropulu's hope for being thrown into a "good loop" by the unpredicted, Stevens's preference for imagination over reason, and the public's belief in the supernatural are not all exactly the same thing. But they are related. A desire for the strange and the surprising seems to be ingrained in human nature. Placed alongside an equal desire for the familiar, the orderly, the rational, we have yet another example of the yin-yang of Chinese philosophy. Literally, shadow and light. Seemingly contrary forces that complement each other and underlie all existence in the natural world. Hot and cold. Low and high. Water and fire. Order and disorder. Rational and irrational.

Nowhere is our ambivalence toward the lawfulness and logic of science more apparent than in our attitudes toward our own bodies and minds. A question that has haunted the discipline of biology since its beginnings—and has not been put entirely to rest in some quarters today—is whether living matter obeys different laws from nonliving matter. As discussed in "The Gargantuan Universe," the "vitalists" have argued that there is a special quality of life—some immaterial or spiritual or transcendent force—that enables a jumble of tissues and chemicals to vibrate with life. That transcendent force is beyond physical explanation. The "mechanists," on the other hand, believe that all of the workings of a living animal can be ultimately understood in terms of the laws of physics and chemistry. Lucretius was a mechanist. Plato and Aristotle were vitalists. They believed that an idealized "final cause," which was more spirit than matter, impelled a germ cell to develop toward an adult form. René Descartes, who famously articulated the separation between the intangible mind and the tangible body, proposed that the immaterial soul interacts with the material body in the pineal gland. In his *Lärbok i kemien*, the most authoritative chemistry textbook of the mid-nineteenth century, Jöns Jacob Berzelius wrote simply: "In living nature the elements seem to obey entirely different laws than they do in the dead."

At the same time as Berzelius's great book, the

mechanists concluded that the energy requirements of animals were supplied solely by the chemical breakdown of food, with no need of a weightless and immaterial spirit or of special laws of nature. (How reminiscent of Laplace's reply to Napoleon.) Still, many members of *Homo sapiens* remained unhappy. The thought that a human body could be reduced to so many coiled springs, balls in motion, weights, and cantilevers has not sat well with many of us.

And what of our minds? Are not our minds merely brains—collections of gooey neurons that store and pass along information in the form of chemicals and electrical ticks, all subject to Coulomb's law and the other mandates of science? Taking the laws of nature and the physicality of the world to their natural conclusion, shouldn't our thoughts and behavior be completely predictable given a large enough computer? If so, then there should be no such thing as irrational behavior. Everything that we think, everything that we say and do in the future, should follow inexorably from the past condition of our brain and the grinding on of the laws.

No, no, no! shrieks the unnamed narrator of Dostoevsky's *Notes from Underground*. In this short novel, one of the first modern literary explorations of the contradictory nature of the mind, the narrator rails against the reason of the intellectual establishment:

[T]his gentleman will at once expound to you, with great eloquence and clarity, precisely how he must needs act in accordance with the laws of reason and truth . . . and then, exactly a quarter of an hour later, without any sudden, extraneous cause, but precisely because of something within him that is stronger than all his interests, he'll cut quite a different caper, that is, go obviously against what he himself was just saying: against the laws of reason, against his own profit; well, in short, against everything . . . He will [do anything to] indeed satisfy himself that he is a man and not a piano key! . . . And more than that: even if it should indeed turn out that he is a piano key, if it were even proved to him mathematically and by natural science, he would still not come to reason, but would do something contrary on purpose, solely out of ingratitude alone; essentially to have his own way.

We will have freedom at any cost. We delight in discovering a rational universe as long as we ourselves are exempt from the rules. We worship order and rationality, but we also have a fondness for disorder and irrationality. I can imagine a futuristic "mind-body" experiment: An intelligent person is placed in a soundproof and sealed room, with minimal sensory input from the external world, and asked a series of questions concerning emotional, aesthetic, and ethical issues. Difficult questions. Suppose also that before entering the room, our test subject's brain is completely examined so that the chemical and electrical state of each neuron is measured and recorded, something that in principle could be done. Then, the puzzle is: Given a very large computer and the known laws of nature, can we predict the person's answer to each of the questions?

Although I am a scientist myself, I would hope not. I cannot explain exactly why. I do believe that the physical universe is governed completely by rational laws, and I also do believe that the body and mind are purely physical. Furthermore, I don't believe in miracles or the supernatural. But, like Dostoevsky's character, I cannot bear the thought that I am simply a piano key, thinking and doing what I must when I'm struck. I want some kind of unpredictability in my behavior. I want freedom. I want some kind of "I-ness" in my brain that is more than the sum of neurons and sodium gates and acetylcholine molecules, a captain who can make decisions on the spot—good or bad decisions, it doesn't matter. Finally, I believe in the power of the mysterious. Einstein once wrote, "The most beautiful experience we can have is the mysterious. It is the fundamental emotion which stands at the cradle of true art and true science." I believe that it is bracing and vital to live in a world in which we do not know all the answers. I believe that we are inspired and goaded on by what we don't understand. And I hope that there will always be an edge between the known and the unknown, beyond which lies strangeness and unpredictability and life.

The Disembodied Universe

In the wee hours of January 8, 1851, working in the cellar of his house on rue d'Assas, a short distance from the Luxembourg Gardens, a small and fragile-looking man named Léon Foucault gave the first direct proof that the Earth spins on its axis. People had waited two thousand years for Foucault's result. Ever since the third century BC, a handful of rebellious thinkers had speculated that the daily sweep of the sun and the stars was caused by the rotation of the Earth, rather than the prevailing view that the heavens revolved about a motionless Earth. But the idea of a spinning Earth had been rejected as a preposterous violation of common sense. After all, we do not live in a state of constant dizziness or feel a cosmic velocity as we hurtle through space. The air does not rush past our ears when we step from our homes. A simple calculation shows that if the Earth rotated once a day on its axis, as some people claimed, the speed of a man standing at the equator would be a staggering 1,000 miles per hour. Aristotle convincingly argued that if the Earth were indeed rotating in an easterly direction, a projectile launched upward would land far to the west. Likewise, clouds and birds would veer to the west. None of these happenings were observed.

Later scientists, however, argued that *if* the Earth spun on its axis, a projectile shot vertically upward would share the land's sideways motion and thus fall in the same spot from which it was launched. And *if* the Earth spun on its axis, the air (and clouds and birds) would be carried along with it and not left behind. Eventually, the new astronomical model hypothesized by Copernicus—in which the Earth orbits the sun and also spins on its axis—gained acceptance by most people. But direct evidence was still not to be had.

In his basement, Foucault hung a twelve-pound brass bob from a six-foot steel wire and set it swinging. A pendulum. His journal from that period reads:

Friday [January 3, 1851] 1–2am First trial, encouraging result. The wire breaks . . .
Wednesday [January 8, 1851] 2am The pendulum turned in the direction of the diurnal motion of the celestial sphere.

The turning of the pendulum was the clincher. Everything depends on your frame of reference. Physicists had shown that the swing of a pendulum remains in the same plane (i.e., does not turn) relative to a *nonrotating* frame of reference. When Foucault's pendulum began slowly turning relative to his laboratory table, which was fixed to the Earth, it could mean only one thing: the Earth was not a nonrotating frame of reference. The Earth rotated. The effect was not large. Every ten minutes, the pendulum turned less than two degrees. But with a sturdy pendulum, not much diminished by friction, and a careful observer, the effect could be measured. Foucault, five feet five inches tall and dismissed as "soft, timid, and puny" by his acquaintances, had originally trained to be a physician but abandoned medicine because he could not stand the sight of blood. Now thirty years old, he was well on his way to becoming one of the great experimental physicists of Europe.

Timid Foucault decided to make a splash with his discovery by mounting a grand demonstration in public. "You are invited to see the Earth turn, in the Meridian Room of the Paris Observatory, tomorrow, from 2pm to 3pm," read a notice he sent out in February. A journalist who attended the performance wrote in *Le National* newspaper: "At the appointed hour, I was there, in the Meridian Room, and I saw the Earth turn."

Of course, the journalist and his friends did not see the Earth turn.

Nor did they feel the Earth turn. Nor did they hear the Earth turn. The turning of the Earth was invisible, and it was silent. It was, in fact, completely hidden to human sensory perceptions. The spectators were informed of this profound but invisible aspect of the world through Foucault's pendulum and their intellectual deductions. Foucault's pendulum, along with the first microscope two hundred years earlier, marked the beginning of a new era in the history of human civilization, in which our knowledge of nature arises not from our own sensory experience but from instruments and calculations. Since Foucault, more and more of what we know about the universe is undetected and undetectable by our bodies. What we see with our eyes, what we hear with our ears, what we feel with our fingertips, is only a tiny sliver of reality. Little by little, using artificial devices, we have uncovered a hidden reality. It is often a reality that violates common sense. It is often a reality strange to our bodies. It is a reality that forces us to re-examine our most basic concepts of how the world works. And it is a reality that discounts the present moment and our immediate experience of the world.

. . .

The most literal discovery of a world beyond human sensory perception was the finding that there is a vast amount of light not visible to the eye. In the mid-nineteenth century, the Scottish physicist James Clerk Maxwell completed a set of four equations that described all electrical and magnetic phenomena. Appropriate manipulations of those equations soon led to other equations that predicted waves moving through space, a bit like water waves moving through water. In Maxwell's case, however, the hypothetical waves were composed of oscillating electrical and magnetic forces instead of crests of water. And the speed of these "electromagnetic waves," a number that came out of the equations, was 186,000 miles per second, the same number previously observed for the speed of light. From the equality of these speeds, Maxwell inferred that the phenomenon we call light is, in fact, a traveling wave of electromagnetic energy. Furthermore, according to the equations, such waves should occur in an enormous range of wavelengths, called the electromagnetic spectrum, from wavelengths much smaller than what the eye can see to wavelengths much larger than what the eye can see. All of these conclusions were hypothetical, mathematical symbols scrawled on pieces of paper. But throughout the history of science, we have learned to take such mathematical calculations seriously. They often describe reality, whether we can see it or not.

One person who took Maxwell's equations seriously was the German physicist Heinrich Hertz. Hertz built an apparatus with an oscillating electrical current—which, according to the theory of Maxwell, should generate electromagnetic waves. This device was his "transmitter." Then Hertz constructed a second device, a "receiver," which consisted of a strand of wire looped around so that its two ends almost touched. Hertz activated his transmitter and placed his receiver on the other side of the lecture hall at the Karlsruhe Physical Institute in Kiel, where he was a professor. Peering carefully at the receiver, he observed faint sparks of electricity jumping across the gap between the two ends of the wire when the transmitter was turned on. Yet he could see nothing but air and stray students in the space between transmitter and receiver. Evidently, as Maxwell had predicted, an invisible wave of energy was traveling from transmitter to receiver. And Hertz could calculate its wavelength, far longer than that of visible light. These invisible waves were radio waves, the first ever produced by human beings. But completely invisible to the human eye. Hertz said to a colleague: "It's of no use whatsoever . . . this is just an experiment that proves Maestro Maxwell was right—

we have these mysterious electromagnetic waves that we cannot see with the naked eye. But they are there."

We now understand that different wavelengths of light correspond to different colors as interpreted by the human brain. The color range visible to the human eye extends from blue light, with a wavelength of about four-hundred-thousandths of a centimeter, to red light, with a wavelength of about eight-hundred-thousandths of a centimeter. But there is a continent of colors redder than red and another continent bluer than blue. Since the time of Maxwell and Hertz, we have built instruments that have detected light of wavelengths several *trillion* times longer than what the eye can see. These are the ultra-long radio waves used for secret communication by submarines. And we have built instruments that have detected light of wavelengths *ten thousand trillion* times shorter than what the eye can see. These are the ultra-high-energy gamma rays produced in the intense gravity of collapsed stars called neutron stars. And all wavelengths between. The portion of the full electromagnetic spectrum visible to the human eye is minuscule. All of these other wavelengths of light are constantly careening through space, flying past our bodies, and presenting strange pictures of the objects that made them—the glow of a warm desert at night, the radio emission of electrons spiraling in the Earth's magnetic field, the X-rays from magnetic storms on the sun. All phenomena invisible to our eyes. But our instruments can see them.

In some ways, we are like the creatures living in Edwin A. Abbott's 1884 novel, *Flatland,* a world of only two dimensions, a world with length and width but no height. Workmen in Flatland are triangles, professional men squares. Priests are circles. The houses in Flatland are pentagons. Rain slides sideways across the two-dimensional sheet of the world, striking shingled roofs, which are straight lines. Life seems fulfilling and complete to the inhabitants of Flatland. They have no conception of a third dimension. Then one day, a visitor from the third dimension arrives. He explains the beauty and richness of his world. The Flatlanders nod their two-dimensional heads, they listen, but they cannot understand. It is the same with our instruments. They inform us of a world far beyond our experience.

In 1905, a German patent clerk named Albert Einstein proposed that our notion of time—a feature of existence so fundamental that it went unquestioned for all recorded history—was in error. Einstein claimed that time was not absolute, that the amount of time elapsed between two events depended on the relative motion of the observers of those events. Einstein did not only suggest. Based on his study of light and a few philosophical principles, he offered a set of equations that precisely quantified how the ticking rates of clocks would differ, depending on their speed relative to each other. For example, 1 second on your clock will be 0.9999999999990 seconds on an identical clock speeding past your clock at 1,000 miles per hour. As can be seen from this example, the discrepancies are tiny at the small speeds familiar in daily life, a fact explaining why no human beings prior to Einstein doubted that a second is a second. But our instruments can measure such small discrepancies and have, in fact, confirmed Einstein's theory. Furthermore, our giant particle accelerators have produced subatomic particles traveling at nearly the speed of light, where "time dilation" is large. A second on your clock is only 0.014 seconds to a particle traveling past you at 99.99 percent of the speed of light. If we were able to move about at such high speeds, time would have a completely different meaning to us. We would constantly need to reset our watches after journeys. When we made a high-speed trip, our children might be older than we were when we returned. When it comes to our bodily experience of time, we are Flatlanders, unable to fathom Einstein's world of relativity.

It is not only modern physics that has uncovered an invisible universe.

Twentieth-century biology has isolated and identified many of the cellular and molecular structures that transmit nervous impulses, store information, control vision and hearing—all far smaller than what can be seen by the eye. Most dramatically, we have discovered the particular molecules that encode the instructions for making new human beings. Each of our trillions of cells, invisible to us except through a microscope, has a complete set of such instructions. What would it feel like if we could see individual molecules, be aware of the trillions of biochemical reactions that take place every second within our bodies, notice when each adenosine triphosphate molecule released a bit of energy to power a muscle, when each neuron in the cerebral cortex went into electrical spasm, when each retinene molecule in the eye straightened out and then twisted again? We are like captains of ships, sitting high on the bridge, who are told about the cabins and engine rooms down below but are never able to see for ourselves.

Perhaps the most startling discovery of a reality beyond sensory perception is that all matter behaves both like particles and like waves. A particle, such as a grain of sand, occupies only one location at each moment of time. By contrast, a wave, such as a water wave, is spread out; it occupies many locations at once. All of our sensory experience with the world tells us that a material thing must be either a particle or a wave, but not both. However, experiments in the first half of the twentieth century conclusively showed that all matter has a "wave-particle duality," sometimes acting like a particle and sometimes acting as a wave.

Evidently, our impression that solid matter can be localized, that it occupies only one position at a time, is erroneous. The reason that we have not noticed the "wavy" behavior of matter is because such behavior is pronounced only at the small sizes of atoms. At the relatively large sizes of our bodies and other objects that we can see and touch, the wavy behavior of particles is only a tiny effect. But if we were subatomic in size, we would realize that we and all other objects do not exist at one place at a time but instead are spread out in a haze of simultaneous existences at many places at once.

The area of science that deals with the wave-particle duality of nature is called quantum physics. The equations of quantum physics and the instruments that have confirmed those equations have revealed a reality that is almost unfathomable from our common understanding of the world. Subatomic particles can be many places at once. Subatomic particles can suddenly disappear from one place and appear in another. And the observer cannot be separated from the observed. The manner in which a particle is observed, in fact, determines the nature of the particle. The world of the quantum is so foreign to our sensory perceptions that we do not even have words to describe it. As Niels Bohr, one of the great figures of modern physics, wrote in 1928: "We find ourselves here on the very path taken by Einstein of adapting our modes of perception borrowed from the sensations to the gradually deepening knowledge of the laws of nature. The hindrances met with on this path originate above all in the fact that . . . every word in the language refers to our ordinary perceptions."

It is an irony to me that the same science and technology that have brought us closer to nature by revealing these invisible worlds have also separated us from nature and from ourselves. Much of our contact with the world today is not an immediate, direct experience, but is instead mediated by various artificial devices such as televisions, cell phones, iPads, chat rooms, and mind-altering drugs. Although few of us know or care about the wave-particle duality of the quantum world, quantum mechanics is, in fact, the science behind the transistor, the computer chip, and all of the modern digital technologies dependent on those devices. Similarly, all of our invisible broadcasts and

receptions from telephone stations, cell phone towers, and wireless modems occur through the invisible
electromagnetic radiation discovered by Maxwell and
Hertz.

But the psychological change accompanying these
technologies is more subtle, and perhaps more important. Consciously and unconsciously, we have gradually
grown accustomed to experiencing the world through
disembodied machines and instruments. As I stood in
line to board an airplane recently, the young woman in
front of me was primping in her mirror—straightening
her hair, putting on lipstick, patting her checks with
blush—a female ritual that has been repeated for several thousand years. In this case, however, her "mirror"
was an iPhone in video mode, pointed at herself, and
she was reacting to a digitized image of herself.

I take walks in a federally protected wildlife preserve near my home in Massachusetts. A dirt trail
winds for a mile around a lake teeming with beavers
and fish, wild ducks and geese, aquatic frogs. Bulrushes
and cattails wrap the perimeter of the pond, water lil-
ies float here and there, rippling when a fish goes by. In
the winter, the air is crisp and sharp, in the summer soft
and aromatic. And a thick silence lies across the park,
broken only by the honking of geese and the croaking
of frogs. It is a place to smell, to see, to feel, to qui-

etly let one's mind wander where it wants. More and more commonly, I see people here talking on their cell phones as they walk around the trail. Their attention is focused not on the scene in front of them, but on a disembodied voice coming from a small box. And they are disembodied themselves. Where are their minds and bodies? Certainly not present in the park. Nor can they be located in the electromagnetic waves and digital signals flowing through cyberspace. Only their voices can be found at the other end of their conversations, in the offices and boardrooms and homes of the people they are talking to. They are attempting to be several places at once, like quantum waves. But I would argue that they are nowhere.

Speaking on the telephone while walking through a nature preserve represents a certain level of disconnection from one's immediate surroundings, but sending text messages is an even greater abstraction. And text messaging is becoming the preferred means of communication by a large segment of the population. In a Nielsen mobile phone survey completed in September 2008, the number of phone calls made by Americans from mid-2006 to mid-2008 remained nearly constant, while the number of text messages increased by 450 percent. The huge increase in text messaging has been driven mostly by teenagers, who have grown up since birth with cell phones and the Internet. According to a Pew survey in 2011, the average American teenager sends or receives 110 text messages a day. When young people go to parks, they are often so busy clicking photos with their iPhones and e-mailing the pictures to their Facebook pages that they do not remember to stop for a moment and contemplate the scene with their own eyes. The most unfortunate aspect of this new behavior is that more and more people, and especially young people, are taking such mediated experiences as "natural," as the norm.

In her 1995 book, *Life on the Screen,* the MIT psychologist and social scientist Sherry Turkle described the way that virtual reality, in the form of "multi-universe domains" and "chat rooms" on the Internet, was beginning to take the place of authentic, face-to-face relationships between people. Many of the younger generation refer to real life as RL and often prefer life on the screen to RL. Turkle goes further in her new book *Alone Together,* where she documents the way in which e-mail and cell phones have created emotional dislocations and superficial but expedient ways to deal with the frantically paced world of the twenty-first century. Leonara, a fifty-seven-year-old chemistry professor in Turkle's study, says: "I use e-mail to make appointments to see friends, but I am so busy that I'm often making an appointment one or two months in the future. After we set things by e-mail, we do not call. I don't call. They don't call. What do I feel? I feel I have 'taken care of that person.'" Audrey, a sixteen-year-old high school student, told Turkle: "Making an [online] avatar and texting. Pretty much the same. . . . You're creating your own little ideal person and sending it out. . . . You can write anything about yourself; these people don't know. You can create who you want to be. . . . Maybe in real life it won't work for you, you can't pull it off. But you can pull it off on the Internet."

All of these examples are now familiar to us. But they are alarming nonetheless. Using technology, we have redefined ourselves in such a way that our immediate surroundings and relationships, our immediate sensory perceptions of the world, are much diminished in relevance. We have trained ourselves not to be present. We have extended our bodies, created enhanced selves that might be called our "techno-selves." Our techno-selves are both bigger and smaller than our former selves. Bigger in that we have tremendous powers to communicate with the invisible world. Smaller in that we have sacrificed some of our contact and experience with the visible, immediate world. We have marginalized our direct sensory experience.

Much of this is an old story, of course. The Romantics of the eighteenth century were rebelling, in part, against the Industrial Revolution and its mechanization of life. Likewise the Hudson River School painters of the mid-nineteenth century, who attempted through their art to recapture the awe and intimacy of a vanishing natural landscape. In Thomas Cole's *River in the Catskills,* for example, a human figure in the foreground looks out upon a peaceful scene of sunlit river, rolling green hills, and faint magenta mountains in the distance. His relaxed pose is a metaphor for the idealized, easy kinship of human being and nature. And the Transcendentalists: "We do not ride the railroad," wrote Thoreau, "it rides upon us."

Since Thoreau, the pace of technological life has increased exponentially, as well as the trend toward an increasingly disembodied experience of the world. The twentieth-century digital technologies have certainly helped enable our techno-selves. But the more penetrating development has been the gradual psychological adaptation to a disembodied experience of the world. When so much of our interaction with other people and with our environment is mediated by the invisible, the visible seems less worthy of our attention. Why should we drive an hour to visit a friend when we can make a Skype call without leaving our house? Or, even more convenient, send a text message? Why should we stare closely at the stippled skin of a snake when we can take a high-resolution digital photograph and magnify the image by ten? In fact, the visible can lead us astray, presenting to us what we consider an inferior reality. We may even be led to mistrust the perceptions and knowledge of our own bodies, in the way airplane pilots are taught to sometimes ignore the sensations of their bodies and rely on their instruments.

When I recently went out to dinner with my twenty-five-year-old daughter and her friends, most of the young women kept their iPhones on the table beside their plates, like miniature oxygen tanks carried everywhere by emphysema patients. Every minute or two one of them glanced down at her device to see what new messages had arrived and to send out other messages. One of the young women showed the others a digitized photograph of her dog. Another played music from her iPod. Occasionally, a factual question would come up as they talked. Conversation stopped, while somebody went on the Internet and looked up the answer. This disembodied existence is their reality. This relation to the world is for them the natural order of things. I myself did not feel like I was sitting at a table with my daughter and her friends, as I did ten or fifteen years ago. I felt like I had been digitized myself, that we were all megabytes being streamed through the Web. Spoken words and facial expressions were just two channels among many.

I would not attempt to argue that the deepening scientific knowledge of the invisible world—the spin of the Earth, the X-rays and radio waves, the dilation of time, the wavy nature of subatomic particles—has directly led to our disembodied life in the world of today. But I would argue that this knowledge, not to mention the technology that has emerged from it, has created a working familiarity with the invisible. And that familiarity, in turn, helps to de-emphasize the vitality of the visible and the directly experienced world. A young child learns that she can press the button of a remote and the picture on the television changes. Or she can go to her father's computer screen and see her mother, a thousand miles away.

As these trends toward disembodied existence continue, it is hard to imagine the world a hundred years from now, just as people living a hundred years ago could not have imagined the world of today. My guess is that a century from now, we will be part human and part machine. We might have electronic ears, we might have special lenses in our eyes that can see X-rays and gamma rays. Twenty-second-century iPhones might create laser holograms of correspondents, so that we can see 3-D moving pictures of distant people as we converse with them. We might have computer chips implanted directly into our brains, so that we have instant access to the galaxies of information on the Internet. Such computer chips, attached to our neurons, might allow us to learn a new language in five seconds, experience memories of events that never actually happened, feel the sensations of sex while sitting alone in a chair. By pushing a button in our twenty-second-century homes, we will fill the room with artificial smells of peonies and lavender, summer grass, fresh-baked bread. Another button will create a hologram of mountains and trees, the nature preserve where I go walking.

Most of us will adapt to this new way of living the same way that the people of today have adapted to cell phones and Skype. It will be the natural and normal way of being in the world. But here and there, small pockets of people will rebel and establish protected communes, where the newer technologies are left at the front gate—in the same way that some people today still send handwritten letters and take long walks without their cell phones. In such enclaves, people will feel that they have preserved something of value, that they are living a more immediate and authentic life, that they are more connected with themselves and their surroundings. And that will be partly true. Yet they will be also disconnected from the larger world just outside of their gate, invisible in their own way.

ACKNOWLEDGMENTS

I would like to thank a number of magazine editors who encouraged me in the writing of several of these essays: Christopher Cox at *Harper's,* Kerry Lauerman at *Salon,* and Cheston Knapp at *Tin House.* For cheerfully allowing themselves to be subjects, I thank Alan Guth, Steven Weinberg, Owen Gingerich, and Garth Illingworth. My longtime editor at Pantheon, Dan Frank, suggested several of these essays and has always encouraged and supported my work. I am eternally grateful to my longtime literary agent, Jane Gelfman.

Finally, I thank my longtime wife, Jean, who bravely offers herself as the first reader of everything I write.

NOTES

THE ACCIDENTAL UNIVERSE

5 "The multiple universe idea severely limits": Comments by Alan Guth made in interviews with the author on May 9, 2011, and July 28, 2011.

5 "We now find ourselves at a historic fork": Comments by Steven Weinberg made in interview with the author on July 28, 2011.

12 "To get our universe": Francis Collins, 31st annual Christian Scholars' Conference at Pepperdine University, June 16, 2011, quoted in the *Christian Post,* June 21, 2011.

15 "This is not your father's universe": Robert Kirshner, National Science Foundation (NSF) Symposium, "Ground Based Astronomy in the 21st Century," Omni Shoreham Hotel, Washington, DC, October 7–8, 2003.

THE TEMPORARY UNIVERSE

30 a photograph of the coast near Pacifica, California: Photograph of Pacifica, California, available at http://miraimages.photoshelter.com/image/I0000dJXI5vwD7QQ.

31 "But I am constant as the northern star": William Shakespeare, *Julius Caesar*, III, i, 60–62.

32 According to astrophysical calculations: For the astrophysical calculations of very long time scales in the universe, see Freeman Dyson, "Time Without End," *Reviews of Modern Physics* 51, no. 3 (July 1979): 447–60.

33 "Impermanent are all component things": *Digha Nikaya, Mahaparinibbana Sutta,* trans. Sister Vajira and Francis Story (Kandy, Sri Lanka: Buddhist Publications Society, 1998), p.16.

35 "A man can do what he wants": Arthur Schopenhauer, *On the Freedom of the Will* (1839): "Der Mensch kann tun was er will; er kann aber nicht wollen was er will."

THE SPIRITUAL UNIVERSE

PART I

38 "Theater has always been about religion": Comments by Alan Brody made in interview with the author, July 10, 2011.

41 See, for example, *God's Activity in the World: God's Activity in the World: The Contemporary Problem,* ed. Owen Thomas (Chico, CA: Scholars Press, 1983).

42 see, for example, Charles Hodge: Charles Hodge, *Systematic Theology,* 3 vols. (1871–73; Peabody, MA: Hendrickson Publishers, 1999).

42 A recent study by the Rice University sociologist Elaine Howard Ecklund: Elaine Howard Ecklund, *Science vs.*

Religion: What Scientists Really Think (Oxford: Oxford University Press, 2010).

43 "I've not had a problem": Francis Collins, *Newsweek,* December 20, 2010.

43 "The universe exists because of God's actions": Ian Hutchinson, interview with the author, July 7, 2011.

43 "I believe that our physical universe is somehow wrapped": Owen Gingerich, interview with the author, July 7, 2011.

47 "We should try to love the questions themselves": Rainer Maria Rilke, *Letters to a Young Poet,* trans. M. D. Herter Norton, rev. ed. (New York: W. W. Norton, 1993), letter 4, July 16, 1903.

50 "Faith is the great cop-out": Richard Dawkins, speech at the Edinburgh International Science Festival, April 15, 1992, published in *The Independent,* April 20, 1992.

50 "Many of us saw religion": Richard Dawkins, *The Guardian,* October 11, 2001.

PART II

57 "Were one to characterize religion": William James, *Varieties of Religious Experience* (1902; BiblioBazaar, 2007), p. 60.

58 "I remember the night": Ibid., p. 71.

59 "Our impulsive belief is here always": Ibid., p. 77.

60 as beautifully described in the book *Personal Knowledge*: Michael Polanyi, *Personal Knowledge* (Chicago: University of Chicago Press, 1958).

62 "the rest from Man or Angel": John Milton, *Paradise Lost,* Book VIII. See, for instance, vol. 4 of the Harvard Classics edition (Cambridge, MA, 1909–14), p. 245.

THE SYMMETRICAL UNIVERSE

69   "We're reaching into the fabric of the universe": Joe Incandela, quoted in Paul Rincon, "Higgs Boson-Like Particle Discovery Claimed at LHC," BBC News, July 4, 2012. Available at http://www.bbc.co.uk/news/world-18702455.

72   "Symmetry principles have moved to a new level": Steven Weinberg, *Dreams of a Final Theory* (New York: Pantheon, 1992), pp. 142, 165.

78   Experiments published in 2004: I. Rodriguez et al., "Symmetry Is in the Eye of the Beeholder: Innate Preference for Bilateral Symmetry in Flower-Naive Bumblebees," *Naturwissenschaften* 91 (2004): 374–77.

79   "A sense of beauty has been declared": Charles Darwin, "Sense of Beauty," in *The Descent of Man* (New York: D. Appleton and Company, 1871), p. 61.

81   "However we analyse the difference": E. H. Gombrich, *The Sense of Order*, 2nd ed. (London: Phaidon, 1984), p. 9.

THE GARGANTUAN UNIVERSE

88   "That little red dot is hellishly far away": This and subsequent comments from Garth Illingworth made in interview with the author on February 11, 2012.

88   A clay tablet dating from the twenty-fifth century BC: See James D. Muhly, "Ancient Cartography: Man's Earliest Attempts to Represent His World," available at http://www.penn.museum/documents/publications/expedition/PDFs/20-2/Ancient%20Cartography.pdf.

Also see "History of Cartography," at http://en.wikipedia
.org/wiki/History_of_cartography.

99 "Man is fallen; nature is erect": Ralph Waldo Emerson,
"Nature." See, for instance, vol. 5 of the Harvard Classics
edition (Cambridge, MA, 1909–14), p. 228.

100 The totality of living matter on Earth: Earth's mass is
6×10^{27} grams. The biomass on Earth is about 6×10^{17}
grams. See, for instance, William B. Whitman, David
C. Coleman, and William J. Wiebe, "Prokaryotes: The
Unseen Majority," *Proceedings of the National Academy of
Sciences* 95, no. 12 (1998): 6578–83. To get the fraction
of living mass in the visible cosmos, I assume that our
star is an average star with a mass of 2×10^{33} grams. And
I assume that 3 percent of all stars have habitable planets
attached.

THE LAWFUL UNIVERSE

106 "If a man proceeded by force": O. R. Gurney and S. N.
Kramer, "Two Fragments of Sumerian Laws," *Assyrio-
logical Studies*, no. 16 (April 21, 1965): 13–19. See also "Code
of Ur-Nammu," at http://en.wikipedia.org/wiki/Code
_of_Ur-Nammu.

108 "I personally do not want": Maria Spiropulu, quoted in
Dennis Overbye, "Physicists Find Elusive Particle Seen as
Key to Universe," *New York Times,* July 4, 2012.

110 "This terror of mind": Lucretius, *De rerum natura*, trans.
and ed. W. H. D. Rouse and M. F. Smith, Loeb Classical
Library (Cambridge, MA: Harvard University Press, 1982),
Book I, lines 146–58.

110 "mist and smoke disperse": Ibid., Book III, lines 136–39.

111 "Therefore, death is nothing to us": Ibid., Book III, line 830.

111 "I believe that our physical universe is somehow wrapped": Owen Gingerich, interview with the author, July 7, 2011.

112 "Any body wholly or partially immersed": Archimedes, "On Floating Bodies." See www.archive.org/stream /worksofarchimede00arch#page/256/mode/2up.

112 the Persian physicist Ibn Sahl: See "The First Steps for Learning Optics: Ibn Sahl's, Al-Haytham's and Young's Works on Refraction as Typical Examples," in Mourad Zghal et al., *Education and Training in Optics and Photonics,* OSA Technical Digest series (Optical Society of America, 2007). See http://en.wikipedia.org/wiki/Ibn_Sahl and also http://spie.org/etop/2007/etop07fundamentalsII.pdf.

113 "mere mechanical causes": Isaac Newton, *The Principia,* vol. 2, trans. I. Bernard Cohen et al. (1687; Berkeley: University of California Press, 1999), pp. 544–45.

113 "Motion is much more apt to be lost": Isaac Newton, *Optiks* (1704), Book III, Part 1. See, for instance, *Great Books of the Western World,* vol. 34 (Chicago: University of Chicago Press, 1952), pp. 540 and 542.

114 "Je n'avais pas besoin de cette hypothèse-là": Pierre-Simon Laplace, quoted in Augustus De Morgan, "On Some Philosophical Atheists," in *A Budget of Paradoxes,* vol. 2 (London: Longman, Greens, 1872). See also http://en.wikisource .org/wiki/Budget_of_Paradoxes/J.

116 "Dear Radioactive Ladies and Gentlemen": Pauli Archive, CERN, Geneva, Switzerland. For the original letter, see http://www.library.ethz.ch/exhibit/pauli/neutrino_e

.html and the translation into English at http://www.pp
.rhul.ac.uk/~ptd/TEACHING/PH2510/pauli-letter
.html.

118 In their excellent book *Wonders and the Order of Nature*:
Lorraine Daston and Katharine Park, *Wonders and the
Order of Nature, 1150–1750* (Cambridge, MA: Zone Books,
1998).

118 "the passion of surprise and wonder": David Hume, "Of
Miracles," in *An Enquiry Concerning Human Understanding* (1748). See for example the Harvard Classics edition, vol. 37 (Cambridge, MA: Harvard University Press,
1909–14), p. 404.

118 "Curiosity pleases me": Michel Foucault, *Foucault Live:
Interviews (1961–84)*, trans. John Johnston, ed. Sylvere
Lotinger (New York: Semiotext[e], 1989), pp. 198–99.

119 "It is the mundo of the imagination": Wallace Stevens,
"The Figure of the Youth as Virile Poet," in *The Necessary
Angel: Essays on Reality and the Imagination* (London: Faber
& Faber, 1960), p. 58.

119 Or the recent Pew survey: Pew Research Center Forum
on Religion and Public Life, December 2009, http://www
.pewforum.org/Other-Beliefs-and-Practices/Many
-Americans-Mix-Multiple-Faiths.aspx#5.

120 "In living nature": Jöns Jacob Berzelius, translated and
quoted in Henry M. Leicester, "Berzelius," *Dictionary of
Scientific Biography*, vol. 2 (New York: Scribner's, 1981),
p. 96a.

122 "[T]his gentleman will at once expound to you": Fyodor
Dostoevsky, *Notes from Underground* (1864), trans. Richard Pevear and Larissa Volokhonsky (New York: Vintage,
1993), pp. 21–22, 30–31.

123 "The most beautiful experience": Albert Einstein, "The World as I See It" *Forum and Century* 84 (1931): 193–94; reprinted in Albert Einstein, *Ideas and Opinions* (New York: Modern Library, 1994), p. 11.

THE DISEMBODIED UNIVERSE

125 the first direct proof that the Earth spins on its axis: Prior to Foucault's pendulum, there was only indirect, nonlocal evidence that the Earth rotates. In 1736–37, Pierre-Louis Maupertuis measured the shape of the poles of the Earth and showed that they are flattened relative to a perfect sphere, and in the 1740s, Charles Marie de La Condamine and Pierre Bouguer measured the equatorial regions of Earth and showed that they bulged relative to a perfect sphere. These altered shapes occur when a nonrigid sphere rotates.

126 His journal from that period reads: Foucault's journal entries can be found in William Tobin, *The Life and Science of Léon Foucault* (Cambridge: Cambridge University Press, 2003), p. 139.

127 dismissed as "soft, timid, and puny": See *The Life and Science of Léon Foucault*, pp. 15 and 18, and references therein.

127 "You are invited to see the Earth turn": Terrien, *Le National*, February 19, 1851. See also Tobin, *The Life and Science of Léon Foucault*, p. 141.

127 "At the appointed hour": Terrien, *Le National*, February 19, 1851.

130 "It's of no use whatsoever": David G. Luenberger, *Information Science* (Princeton, NJ: Princeton University Press,

2006), p. 355. See also Heinrich Hertz, *Electric Waves*, trans. D. E. Jones (1900; New York: Dover, 1962).

136 "We find ourselves here on the very path": Niels Bohr, *Nature Supplement,* April 14, 1928.

138 In a Nielsen mobile phone survey: Marguerite Reardon, "Americans Text More Than They Talk," at http://news .cnet.com/8301-1035_3-10048257-94.html.

139 According to a Pew survey: The Pew Research Center's Internet and American Life Project, April 26–May 22, 2011, Spring Tracking Survey.

139 "I use e-mail to make appointments": Leonara in Sherry Turkle, *Alone Together* (New York: Basic Books, 2011), p. 189.

141 "We do not ride the railroad": Henry David Thoreau, "Where I Lived and What I Lived For," in *Walden* (1854; New York: W. W. Norton, 1951), p. 109.

ABOUT THE AUTHOR

Alan Lightman is the author of six novels, two previous collections of essays, a volume of poetry, and several books on science. His work has appeared in *Harper's*, *The Atlantic*, *Granta*, *The New Yorker*, *The New York Review of Books*, and *Nature*, among many other publications. His novel *Einstein's Dreams* was an international best seller, and his novel *The Diagnosis* was a finalist for the National Book Award in fiction. A theoretical physicist as well as a novelist, he has served on the faculties of Harvard and MIT, and was the first person to receive a dual faculty appointment at MIT in science and in the humanities. Lightman is also the founding director of a nonprofit organization, the Harpswell Foundation, which works to empower women in Cambodia. He lives in the Boston area.